SOVEREIGN TERRA

By Sosé Gjelaj and Elitsa Teneva

ACNOWLEDGMENTS

I am deeply thankful to my siblings, to my children, Victoria Elezovic, Eric Elezovic and Michelle Elezovic and to my grandson, Ethan Elezovic. I am grateful to Elitsa Teneva for partaking a deep interest in the evolutionary process on this planet and her persistent endeavor in bringing awareness out into the world. I am profoundly grateful to Mrs. Snejana Teneva and Mr. Yulii Tenev for assisting us with editing "Sovereign Terra." I would also like to extend my gratitude to Mr. Valentin Tenev for designing the cover of "Sovereign Terra," for his highly creative, innovative and beautiful imagination.

By Sosé Gjelaj

I extend my eternal and profound gratitude to Sosé for her divine heart, mind and actions, for opening the portals of Light and Love and for guiding All Creation along the spiral of evolution. I am also deeply thankful to my beautiful family for their genuine and continual support, constructive feedback and for assisting Sosé and I with the editing and cover design of "Sovereign Terra."

By Elitsa Teneva

Our thanks extend to the cited in "Sovereign Terra" scientists-Avatars without whom the writing of "Sovereign Terra" would have not been possible. Our sincere gratitude is directed to all planetary, galactic, universal and multi-universal co-creators of Light and Love who, consciously or unconsciously, assisted us in writing "Sovereign Terra." Last, but not least we are profoundly grateful to Terra for the limitless gifts that she continuously gives to us.

By Sosé Gjelaj and Elitsa Teneva

TABLE OF CONTENTS

Chapter 1

INTRODUCTION

Exquisite soundless rhythm. Was it a pathless maze, a translucent dream, a ghostly past staring in the mirror? Was it metaphors journeying, generating experiences of our own inner alien encounters to disguise the image of the true selves? Or was it an indigenous cantation from an angel of salvation, the naked masks that hung on the walls?

***

The greatest challenge of our human psyche is that we allowed another to take responsibility of our internal transformation. Instead of looking inside for spiritual inspiration, we looked outside for external admiration. We relied on another to make us aware of our spirit. When our magnificence was not recognized, we leaped into unconscious states. We wrapped our consciousness in a blanket of cognitive dissonance. We surrounded our consciousness with archetypes to the degree of dependency. We surrendered our child tender souls into foreign languages. We focused on the mind rather on the infinite. Our beautiful minds were wrapped into the fabric of the matrix to protect the magician inside. No matter how hard we tried, the influence of the external world did not fulfill the emptiness inside. The expression of our true selves remained in the

1

shadow.

***

The desire for proprietorship creates bureaucracy. Supremacy emphasizes on our limitations. We become pregnant with deceit and give birth to pride.

***

Opposing parties are characters of a play.

***

Problems are the manifestations of human beings making choices that are not aligned with the highest good for all. Obstacles were created to perpetuate greed. Solutions to challenges were always accessible. We failed to implement them because truth was made invisible to humanity. To solve a problem, truth must be revealed. We fell into the dreamer's dream without questioning the imagined truth.

***

Superficial solutions for deeply rooted problems.

***

Cause and effect.

***

Interconnectedness.

***

Unconscious intent morphs the organic into inorganic. In her true self, Mother Earth is organic. So are we. We are sacred *geometry and anything foreign to this geometry loses its meaning and essence of itself.*

***

We are magnificent human beings. We are continuously changing and replacing what has been worn away. We do not need approval for who we are and what we are creating. We have the ability to solve the challenges of the mind so our beaming light can enlighten the hypotheses of our own creation. By delving into the narrowest parts of ourselves we are discovering the vastness of the heart. Unity transcends illusion.

***

Our Mother is a Universal Mother.

By Sosé Gjelaj

In other words, and in most general terms, Sosé Gjelaj establishes a theory which proposes that all life is interconnected. The theory further states that humanity is deceived by archetypal structures which create human and environment-based problems in the name of profit and power. Archetypal structures are herein meant to signify governmental and industrial agencies and key players. Superficial solutions are proposed to mask the root of the problem. Truth is deliberately not revealed to humanity. According to Sosé, the awareness of the truth is the first step to implementing effective and lasting solutions. We are the "magicians" who can heal ourselves and planet Earth.

Based on 445 peer-reviewed scientific articles concerning topics such as water, air and soil pollution, climate change, geoengineering, oil industry, nano and genetically

3

modified (GM) technology, radiation, agriculture, animal consumption and hunting, "Sovereign Terra" provides general and specific, direct and inferred support for the theory established by Sosé.

Chapter 2

CONDITIONS IN MOTION

Overview

"Conditions in Motion" introduces industrial contaminants and more specifically trace metals (i.e., arsenic; lead; mercury; cadmium). Health and environmental impact, relevant policies, practices and court cases are presented. The chapter concludes with environmental risks pertaining to the mining industry and a description of a mining-related case study.

2.1 Industrial Contaminants

Soil, air and water act as transmitters of industrial contaminants into the environment. Industrial pollutants could have adverse effects on health, whether the exposure is to one pollutant or to a combination of contaminants (Conko et al., 2013). Food contamination takes place when industrial pollutants are discharged into the environment (Nduka et al., 2007).

2.1.1 Trace Metals

The global demand for metals will continue to accelerate due to urbanization of the world population (Berteaux, 2013). Anthropogenic (i.e., metal smelting; fossil-fuel combustion; biomass burning) events contribute to the release of trace metals into the atmosphere. Examples of trace metals emitted from anthropogenic activities include arsenic (As), lead (Pb), mercury (Hg) and cadmium (Cd). Health-related as well as environmental problems could result from the emission of trace metals (van Zyl et al., 2014).

2.1.1.1 Arsenic (As)

2.1.1.1.1 Origin & Properties

The earth's surface consists of approximately 0.0001% arsenic (As) (Caceres et al., 2010). Arsenic has accumulative and non-biodegradable properties (Ramudzuli & Horn, 2014). Anthropogenic processes contribute to the presence of As in water, soil (Zheng et al., 2013) and the atmosphere. Anthropogenic As emission accounts for approximately 24,000 tons each year (Benbrahim-Tallaa & Waalkes, 2008). Background As levels in soil range from 1 to 40 mg/kg. As is bound to oxygen, iron, carbon and sulfur to form organic and inorganic arsenicals (iAs) (Hughes, 2006). iAs has greater toxicity compared to organic As (Das et al., 2009). Anthropogenic sources of iAs pollute the air, water and soil with iAs (Hughes, 2006).

2.1.1.1.2 Health Effects

Arsenic is a universal carcinogenic pollutant (Zheng et al., 2013). The carcinogenic properties of As were first observed by British physician Hutchinson in 1887. Besides cancer, exposure to As contributes to the development of a number of health conditions including hypertension, ischemic heart disease, peripheral vascular disorders, arteriosclerosis, dermal lesions, liver disease, diabetes and neuropathy (Benbrahim-Tallaa & Waalkes, 2008). Arsenic affects the nervous system and hence increases the risk of mental-related concerns (Ramudzuli & Horn, 2014).

2.1.1.1.2.1 Acute & Chronic Intoxication

Arsenic intoxication could result following acute or long-term exposure to As (Das et al., 2009). Lubin et al. (2008) found that inhalation of higher levels of As over shorter periods of time is more detrimental than inhaling lower As levels over longer periods of time. Acute As poisoning causes vomiting, nausea, "profuse diarrhea" and abdominal pain (Das et al., 2009). Human tumors induced by As have been observed in populations exposed to excessive levels of As (Schuhmacher-Wolz et al., 2009).

Chronic As poisoning poses a threat to 137 million people in 70 countries (Garcia-Prieto et al., 2012). Chronic exposure to As is linked to cognitive and motor deficits in children (George et al., 2014). Long-term As exposure is also associated with an increased risk of lung, kidney, bladder and skin cancer (Conko et al., 2013). Zhang et al. (2007) conducted a study to investigate the association between exposure to As and genetic damage in human beings. Findings revealed that chronic exposure to As may be linked to gene mutations, deletions, chromosome and DNA damage, alterations in DNA repair and synthesis.

2.1.1.1.2.2 Inorganic Arsenic (iAs)

Arsenicosis results from chronic intoxication with iAs (Hernandez-Zavala et al., 2008). The International Agency for Research on Cancer (IARC) and the U.S. Environmental Protection Agency (EPA) classify iAs as group 1 and group A carcinogen, respectively (Hughes, 2006). Inorganic arsenic is most often associated with skin, lung and urinary bladder tumors. Incidences of liver, kidney and prostate cancer resulting from exposure to iAs have also been reported (Benbrahim-Tallaa & Waalkes, 2008). Inorganic arsenic "readily" crosses the placenta, reaches the fetus and contributes to effects "ranging from developmental toxicity to cancer" (Waalkes & Liu, 2008).

The mechanism of carcinogenicity caused by iAs is not presently known. Possible mechanisms include oxidative stress, genotoxicity, inhibition of DNA repair, cocarcinogenesis, tumor promotion, cell proliferation and

DNA methylation (Hughes, 2006). Taurine (2-aminoetylsuphonic acid) is a free amino acid present in many tissues. It protects the organs of the body against oxidative stress and toxicity caused by toxic compounds (Das et al., 2009).

2.1.1.1.3 Diet

The diet is the primary nonoccupational route to As exposure in human beings. Organic As, primarily prevalent in seafood, is the main source of dietary As (Hughes, 2006). Once ingested, As affects almost all organs of the animal and human body (Das et al., 2009). Arsenic accumulates in hair, skin and nails and has the ability to pass into the placenta, breast milk and the brain (Schuhmacher-Wolz et al., 2009). The daily As intake in the United States (U.S.) ranges between 2 and 92 µg/day (Hughes, 2006).

2.1.1.1.3.1 Inorganic Arsenic (iAs)

The majority of the U.S. adult population is exposed to nonoccupational iAs (Hughes, 2006). Inorganic arsenic adversely impacts millions of people in the world, primarily from drinking iAs-contaminated water (Waalkes & Liu, 2008). Inorganic arsenic is "distributed and taken up" by tissue cells (Schuhmacher-Wolz et al., 2009). Grains (74 ng/g) and produce (9 ng/g) contain the highest iAs levels. Such foods comprise 17-24% of total dietary As. The oral median lethal dose (LD 50) of iAs in human beings is 1-2 mg/kg. The iAs oral reference dose (RfD) or the safe dose of regular daily intake with no adverse effect on health is 0.3 µg/kg/day. The daily dietary iAs consumption for an adult in the U.S. varies between 8 and 14 µg/day (Hughes, 2006).

2.1.1.1.3.2 Rice

Six hundred million tons of rice were produced worldwide in 2003 (Jorhem et al., 2008). Irrigation of rice-crop soils with As-polluted groundwater contributes to rice crop losses. Populations relying on a rice diet are most vulnerable to exposure to As from consuming As-contaminated rice (Zheng et al., 2013). While the average raw rice consumption per capita in Europe is 9 g day^{-1}, the average per capita consumption in the Far East reaches 278 g day^{-1} (Jorhem et al., 2008).

2.1.1.1.4 Cattle Dipping

Arsenic-based cattle dipping practices originated in South Africa in 1911 to "combat East Coast Fever" in imported cattle following the South African War in 1902. Such practices were initiated by the Department of Native Affairs. The disease caused the death of 1.5 million cattle. 10 million cattle were dipped in South Africa by 1960 before the disease was finally eradicated. Children assisted in dipping practices, often without wearing skin protection. Compulsory As-based cattle-dipping practices in South Africa continued until 1983 when they were banned due to expressed human health concerns associated to such practices. Arsenic-based cattle-dipping practices were responsible for As-contaminated soils at and in proximity to unfenced non-operational cattle-dipping sites in South Africa. Children were being observed playing around the dipping sites. Countries such as the U.S., Australia and New Zealand continue to implement As-based animal dipping practices (Ramudzuli & Horn, 2014).

2.1.1.2. Lead (Pb)

2.1.1.2.1 Origin & Properties

Lead is the most significant and widespread non-essential environmental pollutant found in soil, air, water and food. Lead has been present in the environment for thousands of years. Its levels have increased drastically in the last 70 years as a result of anthropogenic activities. Lead has been utilized in the manufacturing of acid batteries, gasoline and until recently, paint. Lead concentrations were present in paint at high levels prior to 1960 while gasoline containing Pb was used until 1996. Pesticides containing Pb were widely used from 1910 until the 1950s. Such products release Pb into the environment. It is predicted that the Pb currently present in the environment will remain within circulation for the next 300 to 500 years due to its tendency to accumulate in soils. Environmental Pb concentrations have reached a point of posing risks to human health (Feleafel & Mirdad, 2013).

2.1.1.2.2 Health Effects

2.1.1.2.2.1 Carcinogenic & Mutagenic Properties

Lead exhibits carcinogenic and mutagenic properties (Feleafel & Mirdad, 2013).

2.1.1.2.2.2 Cardiovascular System

Cardiovascular disease is the number one cause of mortality in the Western world (Mullie et al., 2010). It is a known fact that Pb exposure, even at low environmental levels, has an impact on the cardiovascular system (Park et al., 2006). A number of studies demonstrate an association between Pb and blood pressure, left ventricular hypertrophy, and cardiovascular disease mortality (Gump et al., 2008). Lead is responsible for increasing the blood pressure in adults (Feleafel & Mirdad, 2013).

2.1.1.2.2.3 Children

Lead is transferred from the mother to the fetus and the infant during pregnancy and lactation (Feleafel & Mirdad, 2013). Study findings revealed an association between fetal Pb exposure and adverse health effects on neurodevelopment, particularly as pertaining to the first trimester (Hu et al., 2006). Prenatal Pb exposure was further demonstrated to be linked to a significant increase in cortisol reactivity to acute stress at the age of 9.5 (Gump et al., 2008).

The established by the Centers for Disease Control and Prevention (CDC) blood lead level (BLL) level is 10 μg/dL. Findings from meta-analysis reveal that any level of Pb exposure is "potentially detrimental" (Kim et al., 2008). Lead contamination could contribute to impairment of the central nervous system and death in children (Feleafel & Mirdad, 2013).

According to research, BLLs below the current CDC level result in substantial adverse health effects such as behavioral and learning deficits (Kim et al., 2008). Low levels of Pb could cause long-term increase in cortisol levels and subsequent cognitive deficits (Gump et al., 2008). Results from studies conducted in the 1980s in Australia, North America and Europe demonstrated that childhood blood Pb concentrations of at least 10 μg/dL are inversely related to cognitive test score. "A difference of only a few points on an aptitude test prevents many otherwise eligible students from having an opportunity to pursue higher education."

Childhood blood Pb concentrations have declined in the past 3 decades as a result of the removal of Pb from paint and gasoline. 1988-1994 data demonstrated that about 26% of children between the ages of 1 and 5 years had BLLs between 5 and 10 μg/dL in the U.S. The prevalence of BLLs equal or greater than 10 µg/dL among children between the ages of 1 and 5 in the U.S. declined from 77.8% in the late 1970s to 1.6% between 1999 and 2002 (Jusko et al., 2008). According to the 2003-2004 National Health and Nutrition Examination Survey (NHANES), 2.3% of U.S. children aged between 1 and 5 experience BLLs at or above 10 μg/dL. Such findings indicate that more than 500,000 children less than 6 years of age have elevated BLLs at or above the CDC BLL level of 10 μg/dL (Kim et al., 2008).

The majority of children with elevated BLLs and Pb intoxication in the U.S. are Hispanic or African-American. Such children live in low-income households in urban regions of the U.S. Study findings demonstrated that Pb intoxication is widespread among young children in urban U.S. areas (Kemp et al., 2007). It is imperative to screen at-risk children for Pb in a timely manner. The results of a study revealed that only half of the children in "prioritized screening zones" were screened for Pb (Kim et al., 2008).

2.1.1.2.2.4 Animals

Lead is present in the organs and tissues of mammals. Physiological processes are impeded when Pb is contained in the tissues and organs of mammals in higher concentrations (Feleafel & Mirdad, 2013).

2.1.1.2.3 Vegetable Crops

Vegetables are essential food components of a healthy diet. They provide important nutrients, proteins, vitamins, irons and calcium. Deposits of heavy metals in air, water and soil are absorbed by plants. Leafy vegetables have higher Pb concentrations compared to other vegetables. The lack of sufficient research on the hazard of carcinogenic and mutagenic compounds such as Pb, the rapidly growing population which requires greater agricultural production in addition to the lack of practical alternatives for farmers in terms of chemical and water use have contributed to unethical agricultural practices. Sewage water and industrial waste are often used on agricultural areas to grow crops as a socio-economic efficient strategy. However, sewage and waste water irrigation practices are not safe and sustainable. Such waters contain not only Pb, but other heavy metals absorbed by soils as well.

Ethical agricultural practices such as cultivating crops away from highways or industrial zones, avoiding the use of sewage water on crops, adding organic matter to soils to increase its absorption ability and washing vegetables with 1% vinegar prior to consumption are all strategies designed to reduce Pb concentrations in vegetable crops (Feleafel & Mirdad, 2013).

2.1.1.2.4 *PCS Nitrogen, Inc. v. Ashley II of Charleston*

The Comprehensive Environmental Response, Compensation, and Liability Act (CERCLA), established in 1980 in response to substantial health and environmental risks posed by industrial contamination, mandates that potentially responsible past and present land owners contribute to the cost of de-contamination of polluted by industrial activities land property. CERCLA grants bona fide prospective purchaser (BFPP) status and exemption from liability to land property owners provided liability under CERCLA is based entirely on owning the land property. A high standard of due care is required to obtain a BFPP status and subsequently be exempt from liability as a land owner under CERCLA.

Extensive contaminated land is situated in Charleston, South Caroline. The EPA "had designated the land as a Super-fund site." A number of businesses have used the land for a century, contributing to its pollution (i.e., Pb; As), deterioration and high clean-up cost. Ashley II of Charleston, LLC hoped to clean the property prior to its use, recovering de-contamination cost from past land owners responsible for the contamination. The clean-up process was initiated by Ashley II of Charleston, LLC. In *PCS Nitrogen, Inc. v. Ashley II of Charleston*, the U.S. Court of Appeals for the Fourth Circuit held Ashley alongside past land owners responsible for the clean-up of the property. Ashley stopped the clean-up process and

13

subsequent development of the land due to high liability clean-up costs. The court held that Ashley failed to meet a number of the eight criteria needed for BFPP status. It is argued that the court should have exempted Ashley from clean-up liability under CERCLA as Ashley exercised adequate standards of due care when recovering the site. Industries intending to purchase polluted land and re-develop it will "now be held to a strict standard of care, with limited room for minor mistakes," which will result in "a chill in redevelopment projects and the remediation of far fewer contaminated sites." The court's decision will "ultimately hinder cleanup efforts" (Fisher, 2013).

2.1.1.3 Mercury (Hg)

2.1.1.3.1 Origin & Properties

In a number of developing nations, mercury (Hg) is utilized as part of mining for the amalgamation of metals such as gold. Discharged inorganic Hg from mining contributes to reduced biodiversity, bioaccumulation of Hg in fish, human health risks related to direct contact (i.e., handling Hg) and indirect contact (i.e., consuming contaminated fish) (Picado et al., 2010). Mercury is toxic to wildlife, humans and plants and bioaccumulates in the food chain (Bergin et al., 2005).

2.1.1.3.2 Health Effects

Mercury is a carcinogen, mutagen and teratogen (Goldblum et al., 2013). Depending on the chemical form and dose, high Hg levels in the body could exert nephrotoxic and neurotoxic properties (Maserejian et al., 2008). Children exposed to Hg levels lower than 10 ppm have been noted to exhibit neurobehavioral deficits, particularly pertaining to attention, language, and memory scores (Huel et al., 2008). Prenatal exposure to Hg at elevated levels contributes to "widespread damage to the developing brain" (Huel et al., 2008). Mercury

14

contaminants can be transferred from the mother to the developing fetus. "Methyl mercury is reported to counteract the cardioprotective effects and to damage developing fetuses and young children" (Burger, 2002). Poisoning induced by MeHg has been reported to contribute to developmental disorders (Huel et al., 2008).

2.1.1.4 Cadmium (Cd)

2.1.1.4.1 Origin & Properties

Cadmium levels mirror the degree of industrial contamination in the environment. The release of Cd into the environment results from anthropogenic activities. Aside from being emitted into the air, Cd can contaminate soils which in turn contribute to a pollution of groundwater and of local agricultural crops. The crops are then being used as cattle feed which predisposes animal products, especially vulnerable organs such as the kidneys and liver, to Cd contamination. At the top of the food chain is the consumer who ingests Cd-contaminated agricultural crops and animal products (Vromman et al, 2008).

2.1.1.4.2 Health Effects

Cd is a significant environmental toxin resulting in impaired health following chronic exposure (Vahteristo et al., 2003). Positive correlation has been established between cancer risk and Cd intake (Vromman et al, 2008). The provisional tolerable weekly intake (PTWI) level for cadmium is 7 mg kg^{-1} which translates into a maximum daily intake of 60 mg for a 60-kg person (Vahteristo et al., 2003).

2.2 Mining

Mining activities contribute to "dramatic" degradation of rural ecosystems (Picado et al., 2010). Studies have demonstrated that mining contributes to large-scale deforestation, soil erosion, water shortage and contamination, emission of greenhouse gases and smoldering coal fires (Farahani, 2014). Study findings revealed that over 50 years of metal contamination had a significant impact on genetic diversity among wild yellow perches in mining zones, therefore increasing the incidence of mutation (Wang & Zhou, 2013). Human activities such as mining, hydraulic fracturing and waste disposal could contribute to "induced seismicity" or earthquakes of varying magnitudes. The magnitude of induced earthquakes that have taken place since 1929 is between 1.0 and 7.9. Mining is regarded as a "high-risk activity" not only for people working in mines (i.e., miners; engineers), but also for people living in the surrounding environment (Farahani, 2014).

2.2.1 Mine Tailings

Mine tailings are the materials that remain after the processes of ore extraction and beneficiation. Historically, mine tailings were either disposed of at mine tailings disposal sites, discharged into a receiving pond or into the ocean, lakes or streams. In 1995, a total of 700 million kg of metals in mine tailings were disposed on land every year. Vegetation at un-reclaimed mine tailings disposal sites does not grow for tens to hundreds of years. Mine tailings typically exhibit acidic properties. Solutions to mine tailings disposal sites are typically short-lived. "Although it may not be possible to create an ecosystem equivalent to the surrounding uncontaminated area," a longer-term solution is a technique called phytostabilization. Phytostabilization allows for a restoration of soil ecosystems and the subsequent growth of vegetation on sites (Mendez & Maier, 2008).

2.2.2 Acid Mine Drainage (AMD)

Acid mine drainage (AMD) poses environmental risks (Deventer & Cho, 2014). It contributes to highly acidic soils (pH < 4) (Mentis, 2006). "AMD decants into natural wetland systems, and affects natural ecosystems and agricultural lands" (Deventer & Cho, 2014).

2.2.3 Indigenous Communities

"Historically, large-scale resource development near indigenous communities has produced very few benefits for indigenous people, but significant detrimental impacts." Mining has taken place in North Canada, particularly in the Northwest Territories (NWT). Currently, three diamond mines are active in the NWT. Large-scale mineral development has an effect on land use, indigenous lands and the environment (i.e., environmental degradation; loss of land) (Davidson & Hawe, 2012).

2.2.4 *Center for Biological Diversity v. Salazar*

The Colorado River supplies water to people living in California as well as to 1 in 12 Americans for a total population of approximately 27 million Americans. Uranium deposits are located on the edges of Colorado River. Uranium is a toxic mineral that could solve America's dependency on foreign oil, yet it can also pose risks to water systems of the Southwestern United States.

 In *Center for Biological Diversity v. Salazar*, the U.S. Courts of Appeals for the Ninth Circuit determined that the Bureau of Land Management (BLM) was not mandated to approve a new environmental review prior to permitting a uranium mine to resume mining after ceasing production for almost two decades. A new environmental review would have allowed for a consideration of possible environmental risks under current standards. Uranium mining operations were subsequently resumed in 2009. Though complying with pertinent laws, "the court's ruling exposes a dangerous environmental loophole that requires statutory revisions." The ruling will most certainly contribute to contamination-related health risks to people in Los Angeles who consume tap water derived from the adjacent to the uranium mines Colorado River. It is further concluded that courts "cannot continue to be persuaded by the economically motivated mining industry" (Iannella, 2014).

2.2.5 Soma, Manisa, Turkey

At the end of 2013, miners of a coal mine in Soma, Manisa, Turkey, expressed concerns regarding the safety of the mining conditions. Their "demand" to examine the safety of the mine was rejected in the National Assembly of Turkey. Two weeks later, on May 13, 2014, explosion and fire took place at the coal mine. The explosion took place 2 km underground. The fire continued until May 15, 2014. 301 mine workers died as a result of the explosion (Farahani, 2014).

Discussion

A number of conclusions supporting the interconnectedness and problem-profit aspects of Sosé's theory could be drawn from "Conditions in Motion."

First, the trace metal Pb was used in the manufacturing of paint until 1960 and in gasoline until 1996, indicating that it is possible to manufacture paint and gasoline without the use of Pb. It could thus be inferred that the intent of using Pb in paint and gasoline had been to contaminate the environment and thus human health for hundreds of years into the future. As it is going to become apparent when the pharmaceutical industry is introduced in "Corporal Violations," chronic disease is a highly profitable niche for pharmaceutical companies and health care systems. Pb intoxication causes chronic illness which automatically leads to financial gain for the pharmaceutical and health care sectors. Governmental and industrial profit is therefore the deduced incentive for the deliberate Pb contamination of the environment and of human beings.

Similarly to Pb, it could be further induced that the use of toxic to the environment and human health As, Hg and Cd in manufacturing industries is also driven by the motivation to generate profit from human disease directly resulting from environmental exposure to trace metals. Thus, the use of and failure to effectively regulate and monitor toxic metal release into the environment becomes the means through which governmental and industrial profit from human illness is acquired.

The South Africa's As-based cattle dipping practices example further illustrates how environmental problems are manufactured by governments and industries in order to acquire profit from inflicted human disease. Following the insofar train of thought, it could be inferred that the East Coast fever was purposely invented and dissiminated in order to cause environmental contamination and subsequent human disease. Impaired human health again translates into the generation of income for pharmaceutical and health care industries. The unfenced non-operational As dipping sites in South Africa confirm such conclusions. If profit from human disease was not an incentive, children would have been protected and not exposed to a harsh toxin promoted by the South African government itself.

Last, but not least, the mining case study in Soma, Manisa, Turkey, provides additional support for the theory that governments and industries deliberately orchestrate problems similar to the above examples of human disease in order to acquire profit. In light of the governmental refusal to address miners' safety-related concerns (representative of the fact that governments participate in the corporate crime against humanity), it could be concluded that the explosion

19

served as the intended problem upon which profit was precontemplated to be generated. The message being sent to Soma miners (through fear as the medium) could had been that exposing the truth about Soma mining industry's lack of concern for human safety would result in further demise of miners' life. As will be elaborated on in the "GMO Invasion" chapter, public trust is an important aspect of industrial success. If the public began mistrusting the Soma mining enterprise (due to repeated awareness that human life is not valued by the mining corporation), demand for Soma-mining products may had declined. Thus, the orchestrated explosion prevented possible loss of profit. Alternatively and/or parallel to the above inference, the explosion could had served the purpose of hiding the evidence relative to the inadequacy of Soma mining safety precautions thus avoiding potential liability and resulting loss of profit. In either instance, an explosion taking place only two weeks after miners' formal complaint does not appear to had been coincidental. It is very likely that further investigation into the matter will provide more direct support for the conclusion that the explosion or the problem was intentionally orchestrated to prevent loss and ensure the continual flow of profit for the Soma mining industry.

References

Benbrahim-Tallaa, L. & Waalkes, M. P. (2008). Inorganic arsenic and human prostate cancer. *Environmental Health Perspectives, 116*(2), 158-164.

Bergin, M. S., West, J. J., Keating, T. J. & Russell, A. G. (2005). Regional atmospheric pollution and trasnboundary air quality management. *Annual Review of Environment & Resources, 30*, 1-37.

Berteaux, D. (2013). Quebec's large-scale Plan Nord. *Conservation Biology, 27*(2), 242-247.

Burger, J. (2002). Daily consumption of wild fish and game:

Exposures of high end recreationists. *International Journal of Environmental Health Research, 12*, 343-354.

Caceres, D. D., Werlinger, F., Orellana, M., Jara, M., Rocha, R., Alvarado, S. A. & Luis, Q. (2010). Polymorphism of *Glutathione S-Transferase (GST)* variants and its effect on distribution of urinary arsenic species in people exposed to low inorganic arsenic in tap water: An exploratory study. *Archives of Environmental & Occupational Health, 65*(3), 140-147.

Conko, K. M., Landa, E. R., Kolker, A., Kozlov, K., Gibb, H. J., Centeno, J. A. & Panov, B. S. & Panov. Y. B. (2013). Arsenic and mercury in the soils of an industrial city in the Donets Basin, Ukraine. *Soil and Sediment Contamination, 22*, 574-593.

Das, J., Ghosh, J., Manna, P., Sinha, M. & Sil, P. C. (2009). Arsenic-induced oxidative cerebral disorders: Protection by taurine. *Drug and Chemical Toxicology, 32*(2), 93-102.

Davidson, C. M. & Hawe, P. (2012). All That Glitters: Diamond Mining and Tåîchô Youth in Behchokö, Northwest Territories. *Arctic, 65*(2), 214-228.

Deventer, H. V. & Cho, M. A. (2014). Assessing leaf spectral properties of Phragmites australis impacted by acid mine drainage. *South African Journal of Science, 110*(7/8), 71-82.

Farahani, J. V. (2014). Man-made major hazards like earthquake or explosion; case study Turkish mine explosion. *Iranian Journal of Public Health, 43*(10), 1444-1450.

Feleafel, M. N. & Mirdad, Z. M. (2013). Hazard and effects of pollution by lead on vegetable crops. *Journal of Agricultural & Environmental Ethics, 26*, 547-567.

Fisher, K. (2013). The PCS nitrogen case: A chilling effect on prospective contaminated land purchases. *Boston College Environmental Affairs Law Review, 41*, 29-42.

Garcia-Prieto, J. C., Cachaza, J. M., Perez-Galende, P. & Roig, M. G. (2012). Impact of drought on the ecological and chemical status of surface water and on the content of arsenic and fluoride pollutants of groundwater in the province of Salamanca (Western Spain). *Chemistry and Ecology, 28*(6), 545-560.

George, C. M., Sima, L., Arias, M. H. J., Mihalic, J., Cabrera, L. Z., Danz, D., Checkley, W. & Gilman, R. H. (2014). Arsenic exposure in drinking water: An unrecognized health threat in Peru. *Bulletin of the World Health Organization, 92*, 565-572.

Goldblum, D. K., Cora, M. G. & Rak, A. (2013). Assessing ecological risk from mercury in Little Bay estuary: A study from Fort Totten in New York. *Environmental Quality Management, 23*(1), 23-41.

Gump, B. B., Stewart, P., Reihman, J., Lonky, E., Darvill, T., Parsons, P. J. & Granger, D. A. (2008). Low-level prenatal and postnatal blood lead exposure and adrenocortical responses to acute stress in children. *Environmental Health Perspectives, 116*(2), 249-255.

Hernandez-Zavala, A., Valenzuela, O. L., Matousek, T., Drobna, Z., Dedina, J., Garcia-Vargas, G. G., Thomas, D. J., Del Razo, L. M & Styblo, M. (2008). Speciation of arsenic in exfoliated urinary bladder epithelial cells from individuals exposed to arsenic in drinking water. *Environmental Health Perspectives, 116*(12), 1656-1660.

Hu, H., Tellez-Rojo, M., Bellinger, D., Smith, D., Ettinger, A. S., Lamadrid-Figueroa, H., Schwartz, J., Schnaas, L. Mercado-Garcia, A. & Hernandez-Avila, M. (2006). Fetal lead exposure at each stage of pregnancy as a predictor of infant mental development. *Environmental Health Perspectives, 114*(11), 1730-1735.

Huel, G., Sahuquillo, J., Debotte, G., Oury, J.-F. & Takser, L. (2008). Hair mercury negatively correlates with calcium pump activity in human term newborns and their mothers at delivery. *Environmental Health Perspectives, 116*(2), 263-267.

Hughes, M. F. (2006). Biomarkers of exposure: A case study

with inorganic arsenic. *Environmental Health Perspectives, 114*(11), 1790-1796.

Iannella, D. (2014). Arizona's "zombie" uranium mines: Lax regulations threaten Los Angeles tap water. *Boston College Environmental Affairs Law Review, 41*(3), 54-67.

Jorhem, L., Astrand, C., Sundstrom, B., Baxter, M., Stokes, P., Lewis, J. & Grawe, K. P. (2008). Elements in rice from the Swedish market: 1. Cadmium, lead and arsenic (total and inorganic). *Food Additives and Contaminants, 25*(3), 284-292.

Jusko, T. A., Henderson, C. R., Lanphear, B. P., Cory-Slechta, D. A., Parsons, P. J. & Canfield, R. L. (2008). Blood lead concentrations < 10 μg/dL and child intelligence at 6 years of age. *Environmental Health Perspectives, 116*(2), 243-248.

Kemp, F. W., Neti, P. V. S. V., Howell, R. W., Wenger, P., Louria, D. B. & Bogden, J. D. (2007). Elevated blood lead concentrations and vitamin D deficiency in winter and summer in young urban children. *Environmental Health Perspectives, 115*(4), 630-635.

Kim, D., Galeano, M. A. O., Hull, A. & Miranda, M. L. (2008). A framework for widespread replication of a highly spatially resolved childhood lead exposure risk model. *Environmental Health Perspectives, 116*(2), 1735-1739.

Lubin, J. H., Moore, L. E., Fraumeni Jr., J. F. & Cantor, K. P. (2008). Respiratory cancer and inhaled inorganic arsenic in copper smelters workers: A linear relationship with cumulative exposure that increases with concentration. *Environmental Health Perspectives, 116*(2), 1661-1665.

Maserejian, N. N., Trachtenberg, F. L, Assmann, S. F. & Barregard, L. (2008). Dental amalgam exposure and urinary mercury levels in children: The New England Children's Amalgam Trial. *Environmental Health Perspectives, 116*(2), 256-262.

Mendez, M. O. & Maier, R. M. (2008). Phytostabilization of mine tailings in Arid and Semiarid Environments-an emerging remediation technology. *Environmental Health Perspectives, 116*(3), 278-283.

Mentis, M. T. (2006). Restoring native grassland on land disturbed by coal mining on the Eastern Highveld of South Africa. *South African Journal of Science, 102*, 193-197.

Mullie, P., Clarys, P., Hulens, M. & Vansant, G. (2010). Distribution of cardiovascular risk factors in Belgian army men. *Archives of Environmental & Occupational Health, 65*(3), 135-139.

Nduka, J. K. C., Orisakwe, O. E. & Maduawguna, C. A. (2007). Heavy metals other than lead in flaked paints from buildings in Eastern Nigeria. *Toxicology and Industrial Health, 23*, 525-528.

Park, D-K., Bitton, G. & Melker, R. (2006). Microbial inactivation by microwave radiation in the home environment. *Journal of Environmental Health, 69*(5), 17-24.

Picado, F., Mendoza, A., Cuadra, S., Barmen, G., Jakobsson, K. & Bengtsson, G. (2010). Ecological, groundwater, and human health risk assessment in a mining region of Nicaragua. *Risk Analysis, 30*(6), 916-933.

Ramudzuli, M. R. & Horn, A. C. (2014). Arsenic residues in soil at cattle dip tanks in the Vhembe district, Limpopo Province, South Africa. *South African Journal of Science, 110*(7/8), 64-70.

Schuhmacher-Wolz, U., Dieter, H. H., Klein, D. & Schneider, K. (2009). Oral exposure to inorganic arsenic: Evaluation of its carcinogenic and non-carcinogenic effects. *Critical Reviews in Toxicology, 39*(4), 271-298.

Vahteristo, L., Lyytikainen, T., Venalainen, E.-R., Eskola, M., Lindfors, E., Pohjanvirta, R. & Maijala, R. (2003). Cadmium intake of moose hunters in Finland from consumption of moose meat, liver and kidney. *Food Additives and Contaminants, 20*(5), 453-463.

van Zyl, P. G., Beukes, J. P., du Toit, G., Mabaso, D., Hendriks, J.,

Vakkari, V., Tiitta, P., Pienaar, J. J., Kulmala, M. & Laakso, L. (2014). Assessment of atmospheric trace metals in the western Bushveld Igneous Complex, South Africa. *South African Journal of Science, 110*(3/4), 1-11.

Vromman, V., Saegerman, C., Pussemier, L., Huyghebaert, A., De Temmerman, L., Pizzolon, J.-C. & Waegeneers, N. (2008). Cadmium in the food chain near non-ferrous metal production sites. *Food Additives and Contaminants, 25*(3), 293-301.

Waalkes, M. P. & Liu, J. (2008). Early-life arsenic exposure: Methylation capacity and beyond. *Environmental Health Perspectives, 116*(3), A104.

Wang, Y. & Zhou, J. (2013). Endocrine disrupting chemicals in aquatic environments: A potential reason for organism extinction? *Aquatic Ecosystem Health & Management, 16*(1), 88-93.

Zheng, M.-Z., Li, G., Sun, G.-X., Shim, H. & Cai, C. (2013). Differential toxicity and accumulation of inorganic and methylated arsenic in rice. *Plant Soil, 365*, 227-238.

Chapter 3

A SAMPLE OF INORGANIC ELEMENTS

Overview

"A Sample of Inorganic Elements" explores the effects of toxic chemical substances on human health and on the environment, relevant policies and industrial practices. Reviewed chemical compounds include endocrine-disrupting chemicals (EDCs), pentschlorophenol (PCP), asbestos, phthalates, bisphenol-A (BPA), polychlorinated biphenyls (PCBs), polybrominated diphenyl ethers (PBDEs), volatile organic compounds (VOCs), mold, silicia, triclosan and contaminants associated with office exposure.

* * *

The production and distribution of chemical substances in the environment continues to accelerate (Bechi et al., 2010).

3.1 Endocrine-disrupting chemicals (EDCs)

There are approximately 87,000 potential EDCs (der Mude, 2011). EDCs "can be found within a number of products, including pesticides, pharmaceuticals, plastics..." (Wang & Zhou, 2013).

3.1.1 Human Health Effect

Research findings demonstrate adverse health impacts of EDCs on human beings ranging from reproductive abnormalities, infertility and cancer to metabolic disorders such as obesity and type 2 diabetes. All of these disorders affect the endocrine system. The endocrine system consists of hormone-secreting glands which regulate a large number of the body's most significant functions including blood pressure, metabolism, the nervous system and developmental processes. EDCs impact hormone processes in human beings via the disruption of endocrine functions. EDCs disproportionately affect the human developing fetus. "Malfunctioning" hormones in the mother's womb can result in physical abnormalities of the newborn, apparent immediately following birth. A greater risk of cancer, auto-immune disease and depression later in life have also been demonstrated as a consequence of absorption of EDCs during pregnancy.

Small quantities of EDCs are consumed by human beings on a daily basis. Research findings reveal that small doses of EDCs are more dangerous than large doses (der Mude, 2011).

3.1.2 Aquatic Species Health Effect

Some EDCs are "widely dispersed" in aquatic environments, negatively impacting the developmental behavior, endocrine system and reproductive health of aquatic organisms. Aquatic species gradually lose their ability to reproduce and "face the danger of extinction." EDC pollutants could induce mutagenic outcomes and increase the prevalence of extinction. Adaptation to environments high in EDC pollution is "nearly zero" (Wang & Zhou, 2013).

3.2 Pentachlorophenol (PCP)

Pentachlorophenol (PCP) is highly toxic. The health risks associated with PCP exposure include high fever, damage to the organs, immune system and tissues, difficulties breathing and profuse sweating. The EPA's Science Division demonstrated that PCP "poses an 'unacceptable' cancer risk to children."

13 countries have banned PCP. The United States (U.S.) permits its use as a wood preservative. In the U.S., 36 million

wooden poles leach PCP and other contaminants into nearby soils and water. The U.S. court system has not proposed a resolution to PCP discharge from wooden poles in the U.S. (Cooper, 2014).

3.3 Asbestos

Six naturally occurring fibrous minerals fall under the category of asbestos. An overwhelming number of scientists agree on the fact that there is no safe asbestos concentration (Ramazzini, 2010).

3.3.1 Carcinogenic Properties

All asbestos forms are proven human carcinogens. EPA, the IARC of the World Health Organization (WHO), and the National Toxicology Program (NTP) identified asbestos as a human carcinogen more than 20 years ago. Malignant mesothelioma (MM), ovarian, lung, laryngeal and gastrointestinal cancers are some forms of cancer caused by asbestos.

Millions of workers worldwide have been exposed to asbestos. Between 100,000 and 140,000 asbestos-induced cancer deaths in workers take place worldwide each year. 20,000 new cases of lung cancer and 10,000 new cases of mesothelioma are recorded each year in North America, Japan, Australia and Western Europe as a result of exposure to asbestos. "Asbestos cancer victims die painful lingering deaths." Mortality resulting from asbestos cancer is nearly entirely preventable (Ramazzini, 2010).

3.3.1.1 Malignant Mesothelioma (MM)

"Asbestos is the only established causal factor" for the development of MM (Maule et al., 2007). Approximately 38% of MM could be attributed to exposure to environmental asbestos (Tse et al., 2010). Both occupational and nonoccupational exposure to asbestos increase the risk of MM (Ramazzini, 2010). In the 1980s, it was suggested that one-third of all cases of MM in the U.S. are the result of nonoccupational exposure to asbestos (Maule et al., 2007). The latency period between initial asbestos exposure and the onset of MM ranges between 40 and 60 years (Tse et al., 2010). There is a lack of evidence demonstrating a safety threshold level below which the risk of MM is not present (Ramazzini, 2010).

3.3.2 Asbestos Industry

Similarly to the tobacco industry, the asbestos industry has attempted to obscure the association between asbestos and disease by evoking deceptive scientific debate over the role of fiber types, viruses and genetics in the development of MM and lung cancer. The asbestos industry placed the blame of MM on poliovirus vaccines contaminated with the monkey simian virus 40 (SV40) and used in the 1950s and 1960s. The asbestos industries continue to provoke "endless debate on the relative hazards of asbestos." Such controversies "have helped to make the disease experiences and early deaths of asbestos-exposed workers and people in asbestos-contaminated communities invisible and uncompensated, allowing the asbestos industry to escape accountability" (Ramazzini, 2010).

3.3.2.1 Production & Application

More than 2 million tons of asbestos are produced each year worldwide. Russia is the primary manufacturer of asbestos, followed by China, Kazakhstan, Brazil, Canada, Zimbabwe, and Colombia. The 6 leaders of asbestos production worldwide accounted for 96% of global asbestos production in 2007. Russia alone produces 925,000 tons of asbestos per year, the majority of which is exported. Approximately 70% of the world's production of asbestos is used in "countries desperate for industrial growth." China is the greatest consumer of asbestos worldwide, followed by India, Russia, Kazakhstan, Thailand, Ukraine, and Uzbekistan. 90% of worldwide asbestos is used in the manufacture of asbestos-cement (A-C) pipes, sheets, and water storage tanks.

Schools and commercial buildings in developed countries contain substantial quantities of asbestos as a result of past asbestos-based construction practices. Large quantities of asbestos are still used nowadays in construction in developing countries, contributing to the accumulation of dust contaminated with asbestos. The use of asbestos in such materials continues in spite of warnings of high hazard resulting from the great number of people exposed to asbestos dust and the great difficulty of controlling exposure to asbestos, particularly in children, once established in communities (Ramazzini, 2010).

3.3.2.2 Asbestos Ban

A worldwide ban on the mining, manufacture and use of asbestos was proposed as soon as "evidence of the carcinogenicity of asbestos became incontrovertible." Asbestos is currently banned in 52 countries worldwide, which represent less than one-third of all WHO member countries. A much greater number of WHO countries continue to manufacture, use, import and export asbestos and products containing asbestos. A number of countries that have banned other forms of asbestos exempt the "controlled use" of chrysotile asbestos from the ban (Ramazzini, 2010).

3.3.2.3 Chrysotile Asbestos

Chrysotile asbestos comprises 95% of the worldwide use of asbestos. The exemption of chrysotile asbestos has no ground on medical science. Rather, it "reflects the political and economic influence of the asbestos mining and manufacturing industry." Asbestos industries attest that chrysotile is a safe form of asbestos while a large number of studies refute such assertion. "Controlled use" or safe use of chrysotile asbestos does not exist.

"All countries of the world have an obligation to their citizens to join the international endeavor to ban all forms of asbestos...Safer substitutes for asbestos exist, and they have been introduced successfully in many nations." Safer products have replaced a large number of asbestos-based products in countries where asbestos is banned (Ramazzini, 2010).

3.4 Phthalates

Phthalates are used in the manufacture of household, medical and consumer products (Renner, 2007). Thyroid hormones play a critical role in physiologic systems and affect nearly every tissue in the body. Alterations in thyroid hormone concentrations could contribute to a number of adverse health outcomes (Meeker et al., 2007). Meeker et al., (2007) revealed a possible link between exposure to phthalates and altered major thyroid function indicators.

3.5 Bisphenol-A (BPA)

Bisphenol-A (BPA) is a plastic material used in a wide array of products such as water and baby bottles, contact lenses and as a liner of canned foods including baby infant formula. BPA is associated with breast and prostate cancer, reproductive system abnormalities, obesity, early puberty in girls and neurobehavioral dysfunction. Only small amounts of BPA are required to impact hormonal function. A study conducted by the CDC demonstrated that 95% of participants had BPA traces in their urine. A study by the Environmental Working Group found that 9 out of 10 samples of blood originating from the umbilical cord of newborn babies contained BPA concentrations (der Mude, 2011).

3.6 Polychlorinated biphenyls (PCBs)

Polychlorinated biphenyls (PCBs), "industrial chemicals theoretically composed of a mixture of 209 chlorobiphenyls congeners" are a "widespread aquatic contaminant" with carcinogenic properties (Sisman et al., 2007). The IARC describes PCBs as "probable human carcinogens" (Binnington et al., 2014). It has been shown that PCBs affect the human endocrine system as well as thyroid hormone homeostasis (Turyk et al., 2008). The findings of a study documented that infants who were exposed to high levels of PCBs and polychlorinated dibenzofurans in utero "developed irregular calcification of their skull bones" (Hodgson et al., 2008). PCBs impair fish reproduction, may affect sexual development in female fish and may increase fish mortality rates. It has been demonstrated that sub-lethal concentrations of PCBs are "hazardous" to the embryo development of zebrafish.

The production of PCBs has been banned by many countries, yet their presence is still evident in the environment (Sisman et al., 2007). PCBs "still enter the environment through accidental spillage" (Dhooge et al., 2006).

3.6.1 Neurodevelopmental & Neuropsychological Effects

PCBs have neurological and cognitive toxicity properties (Binnington et al., 2014). There is a substantial evidence of a link between exposure to PCBs and subtle neurodevelopmental impairments in neonates, infants and children (Fitzgerald et al., 2008). Human exposure to PCBs during the first years of life has been associated with lower IQ, impaired motor skills, compromised growth, reproductive abnormalities such as declines in fecundity later in life. (Binnington et al., 2014).

The neurotransmitters serotonin and dopamine have an impact on cognition, olfaction and mood. PCBs exert possible effects on dopamine and serotonin. PCBs may contribute to the rapid progression of motor and cognitive dysfunction characterizing normal aging. The association between exposure to low levels of PCBs and neuropsychological condition among adults aged 55-74 was investigated. Findings demonstrated a possible link between low-level PCBs exposure and memory, depression and learning health effects (Fitzgerald et al., 2008). The results of a different study revealed that a combination of PCBs contributed to reduced dopamine levels and interfered with neurotransmitters, which subsequently had an impact on animal behavior, slowing down kinetic movements (Lin et al., 2008a).

3.7 Polybrominated diphenyl ethers (PBDEs)

Polybrominated diphenyl ethers (PBDEs) are presently used as flame retardants in electronic appliances, textiles, furniture and construction products. PBDEs are similar to PCBs in terms of structural and bioaccumulative attributes. In the past 2 decades, PCB levels in human samples have decreased while PBDEs levels have increased (Turyk et al., 2008).

3.8 Methanol

Methanol is used as a fuel or fuel additive, solvent and starting substance for the synthesis of other industrial chemicals. Methanol is presently being developed as fuel cells to generate electricity for small appliances and personal computers. Exposure to elevated levels of methanol could lead to vision impairment or blindness in human beings (Cruzan, 2009).

3.9 Volatile Organic Compounds (VOCs)

Volatile organic compounds (VOCs) are widely used in deodorizers, dry-cleaning processes, degreasing and cleaning products, pesticides, paints, solvents and personal care products. A number of VOCs are also present in industrial emissions and automotive exhaust. VOCs can further be released into the atmosphere during showering in chlorinated water. VOCs "exhibit acute and chronic toxicity in people" (Sexton et al., 2005).

3.10 Mold

Jaakkola et al. (2005) investigated the association between domestic mold exposure and childhood asthma development. The results of the study revealed a positive association between mold exposure and the development of asthma in children.

3.11 Silicia

Silica is a mineral. Crystalline silica dust is often referred to as quartz. Crystalline silica is used in a number of industries such as cement manufacturing, ceramics, clay, china and glass pottery, manufacture of glass and concrete mixing products, electronics, foundry, manufacturing and sand-blasting abrasives and blast furnaces. The IARC classified crystalline silica as a human carcinogen. Inhalation of silica dust represents a health hazard. An association has been established between exposure to crystalline silica and increased risk of rheumatoid arthritis, seleroderma, lupus, renal disease, Sjogern's syndrome, lung cancer and obstructive pulmonary disease (Yassin et al., 2005).

3.11.1 Silicosis

Silicosis is a health condition characterized by the formation of scar tissue in the lungs, reducing the ability to obtain oxygen from air. A large number of occupational workers are exposed to respirable quartz (Yassin et al., 2005). "Silicosis is a major occupational health concern" (Nelson et al., 2010). It has been estimated that approximately 2 million workers in the U.S. are exposed to crystalline silica dust in the general, construction and maritime industries. There were 3,600-7,300 new silicosis cases in the U.S. on an annual basis between 1987 and 1996. Silicosis was identified as the causal factor of 200-300 deaths per year in the U.S. between 1990 and 1996 (Yassin et al., 2005).

The prevalence of silicosis is especially high in developing and mining countries (Nelson et al., 2010). Nelson et al., (2010) analyzed silicosis trends among South African gold miners who underwent autopsy between 1975 and 2007. Results did not reveal a reduction in the incidence of silicosis during the 33-year study period. Findings further suggested that "the burden of silicosis will continue to rise" as gold miners continue aging and working for longer periods of time. The high prevalence of silicosis among gold mine workers in South Africa indicated that gold mining companies were unable to reduce silica dust to levels that were considered safe. A large number of South Africa's gold miners "are already burdened with tuberculosis and/or HIV." A priority of the International Labour Organization and the WHO is to eradicate silicosis.

3.12 Triclosan

Triclosan is an antimicrobial synthetic chemical that has been widely used for more than 2 decades in a number of consumer products such as textiles, plastic kitchenware, personal care products (i.e., toothpaste; mouthwash; soaps; deodorants) and toys. 76% of 395 commercial soaps in the U.S. were found to contain triclosan. Approximately 350 tons of triclosan are produced in Europe for commercial use each year. The efficacy of triclosan in health-care related and household settings as well as the emergence of antibiotic-resistance bacteria associated with its use, "are the subject of an ongoing

35

scientific and public debate."

Triclosan has been found in aquatic settings as well as in certain food sources. It is not acutely toxic to mammals. Studies have demonstrated thyroid-related disrupting effects of triclosan in frogs (Calafat et al., 2008).

3.13 Office Exposure Contaminants

An increasing number of the workforce is employed in indoor office spaces. A few past studies have established an association between office exposure and symptoms of sick building syndrome such as nose, throat, eye and skin adverse health effects, cough and fatigue. A link between carbonless copy paper (CCP) and paper dust exposure and increased risk of onset of asthma had been demonstrated (Jaakkola & Jaakkola, 2007).

Discussion

"A Sample of Inorganic Elements" also stresses on the interconnectedness and problem-profit aspects of Sosé's theory.

Presently, there are 87,000 potential EDCs. Inferring from Sosé's theory, EDCs were intentionally created and disseminated to generate profit from human disease and environmental degradation (i.e., fundraising) associated with exposure to EDCs. Extinction of wild fish populations further leads to the rise of aquaculture which (as it is going to become apparent in the "Fish Exhaustion" chapter) is generally owed by a few large industries. This automatically translates into additional centralized corporate profit.   The interconnectedness between created problems and opportunities for profit is inexhaustible. If a strong incentive such as profit was not present, governments and industries would have not permitted the staggering rates of pollution of the environment with toxic elements such as EDCs.

To add, if profit did not drive the orchestration of problems, all precautionary measures would had been taken to preven the environmental release of PCBs by accident while PCBs would had not been replaced by the equally harmful PBDEs. If profit through problems was not the focus of both industries and governments, asbestos would had been banned worldwide and replaced with its safe alternatives. However, safe asbestos alternatives do not cause MM or lung disease and the resulting from medical treatment profit which is a viable reason for asbestos to be only partially banned and safe asbestos alternatives – partially implemented.

References

Bechi, N., Ietta, F., Romagnoli, R., Jantra, S., Cencini, M., Galassi, G., Serchi, T., Corsi, I., Focardi, S. & Paulesu, L. (2010). Environmental levels of *para*-Nonylphenol are able to affect cytokine secretion in human placenta. *Environmental Health Perspectives, 118*(3), 427-431.

Binnington, M. J., Quinn, C. L., McLachlan, M. S. & Wania, F. (2014). Evaluating the effectiveness of fish consumption advisories: Modeling prenatal, postnatal, and childhood exposures to persistent organic pollutants. *Environmental Health Perspectives, 112*(2), 178-186.

Calafat, A. M., Ye, X., Wong, L.-Y., Reidy, J. A. & Needham, L. L. (2008). Urinary concentrations of triclosan in the U.S. Population: 2003-2004. *Environmental Health Perspectives, 116*(3), 303-307.

Cooper, C. B. (2014). Toxic solid waste leaching from telephone poles? Navigating ambiguous definitions in RCRA. *Environmental Affairs, 41*(14), 14-29.

Cruzan, G. (2009). Assessment of the cancer potential of methanol. *Critical Reviews in Toxicology, 39*(4), 347-363.

der Mude, A. V. (2011). Endocrine-disrupting chemicals: Testing to protect future generations. *Environmental Affairs, 38*, 509-535.

Dhooge, W., van Larebeke, N., Koppen, G., Nelen, V., Schoeters, G., Vlietinck, R., Kaufman, J.-M. & Comhaire, F. (2006). Serum dioxin-like activity is associated with reproductive parameters in young men from the general Flemish population. *Environmental Health Perspectives, 114*(11), 1670-1676.

Fitzgerald, E. F., Belanger, E. E., Gomez, M. I., Cayo, M., McCaffrey, R. J., Seegal, R. F., Jansing, R. L., Hwang, S.-a. & Hicks, H. E. (2008). Polychlorinated biphenyl exposure and neuropsychological status among older residents of Upper Hudson River communities. *Environmental Health Perspectives, 116*(2), 209-215.

Hodgson, S., Thomas, L., Fattore, E., Lind, P. M., Alfven, T., Hellstrom, L., Hakansson, H., Carubelli, G., Fanelli, R. & Jarup, L. (2008). Bone mineral density changes in relation to environmental PCB exposure. *Environmental Health Perspectives, 116*(9), 1162-1166.

Jaakkola, J. J. K., Hwang, B.-F. & Jaakkola, N. (2005). Home dampness and molds, parental atopy, and asthma in childhood: A six-year population-based cohort study. *Environmental Health Perspectives, 113*(3), 357-361.

Jaakkola, M. S., Jaakkola, J. J. K. (2007). Office work exposures and adult-onset asthma. *Environmental Health Perspectives, 115*(7), 1007-1011.

Lin, K.-C., Guo, N.-W., Tsai, P.-C., Yang, C.-Y., Guo, Y.-L. L. (2008a). Neurocognitive changes among elderly exposed to PCBs/PCDFs in Taiwan. *Environmental Health Perspectives, 116*(2), 184-189.

Meeker, J. D., Calafat, A. M. & Hauser, R. (2007). Di(2-ethylhexyl) phthalate metabolites may alter thyroid hormone levels in men. *Environmental Health Perspectives, 115*(7), 1029-1034.

Maule, M. M., Magnani, C., Dalmasso, P., Mirabelli, D., Merletti, F. & Biggeri, A. (2007). Modeling mesothelioma risk associated with environmental asbestos exposure. *Environmental Health Perspectives, 115*(7), 1066-1071.

Nelson, G., Girdler-Brown, B., Ndlovu, N. & Murray, J. (2010). Three decades of silicosis: Disease trends at autopsy in South African gold miners. *Environmental Health Perspectives*, 118(3), 421-426.

Ramazzini, C. (2010). Asbestos is still with us: Repeat call for a universal ban. *Archives of Environmental & Occupational Health*, 65(3), 121-126.

Renner, R. (2007). New phthalate link? DEHP metabolites and altered thyroid hormone levels in men. *Environmental Health Perspectives*, 115(7), A 363.

Sexton, K., Adgate, J. L., Church, T. R., Ashley, D. L., Needham, L. L., Ramachandran, G., Fredrickson, A. L. & Ryan, A. D. (2005). Children's exposure to volatile organic compounds as determined by longitudinal measurements in blood. *Environmental Health Perspectives*, 113(3), 342-349.

Sisman, T., Geyikoglu, F. & Atamanalp, M. (2007). Early life-stage toxicity in zebrafish (Danio rerio) following embryonal exposure to selected polychlorinated biphenylis. *Toxicology and Industrial Health*, 23, 529-536.

Tse, L. A., Yu, I. T-s., Goggins, W., Clements, M., Wang, X. R., Au, J. S.-k. & Yu, K. S. (2010). Are current or future Mesothelioma epidemics in Hong Kong the tragic legacy of uncontrolled use of asbestos in the past? *Environmental Health Perspectives*, 118(3), 382-386.

Turyk, M. E., Persky, V. W., Imm, P., Knobeloch, L., Chatterton Jr., R. & Anderson, H. A. (2008b). Hormone disruption by PBDEs in adult male sport fish Consumers. *Environmental Health Perspectives*, 116(12), 1635-1641.

Wang, Y. & Zhou, J. (2013). Endocrine disrupting chemicals in aquatic environments: A potential reason for organism extinction? *Aquatic Ecosystem Health & Management*, 16(1), 88-93.

Yassin, A., Yebesi, F. & Tingle, R. (2005). Occupational exposure to crystalline silica dust in the United States, 1988-2003. *Environmental Health Perspectives*, 113(3), 255-260.

Chapter 4

LEATHER "PRIVILEGE"

Overview

"Leather 'Privilege'" elaborates on the number of tanneries in major leather-producing countries, industrial processes pertaining to producing leather goods, the environmental damage and water consumption resulting from leather industry processes and child labor in leather-producing settings.

* * *

Leather is defined as "the dressed or tanned hide of an animal, usually with the hair removed" (Syed et al., 2010).

4.1 Pakistan

There are approximately 600 tanneries in Pakistan. The leather sector in Pakistan employs more than 1 million people and is a leading contributor to the national economy. Untreated effluent from leather industries in Pakistan is being increasingly discharged into reservoirs, leading to severe environmental degradation. Three types of waste are produced by leather industries in Pakistan – wastewater, solid waste and air waste. Wastewater represents the most significant environmental challenge. 50 to 60 L of water is required per kilogram of hide. The water contaminated with chemicals used to treat the hide is thereafter discharged into water reservoirs. Discharged wastewater enters the food web and drinking water supplies, posing health risks to humanity. 5,500 kg of solid waste is produced every day from treating 10,000 kg of skin per day.

The Pakistan leather industry also has detrimental

effects on human health. Health risks associated with the leather industry include various kinds of cancer, hematological disorders such as anemia and leukemia, neurological disorders, musculoskeletal impairments and psychological aspects related to the stress of work (Syed et al., 2010).

4.2 North Africa & Brazil

The leather sector in the North African regions (i.e., Morocco, Algeria, Tunisia and Egypt) includes 407 tanneries, 1,000 leather goods businesses and 7,000 footwear companies. There are approximately 8,000 tanneries located in Brazil. Brazil produced 755 million pairs of shoes in 2004. Over 27 million leather goods were exported from Brazil in 2007. Close to 2 million people are employed in the Brazilian tanneries. Brazilian tanneries and footwear industries generate approximately 1,400 tons of waste. The Brazilian leather industry uses 30 to 80 m^3 of water per ton of processed hide. The average water used for a processed hide is 630L. A single leather tannery in Brazil produces 3,000 hides per day, consuming 1900 m^3 of water. Such amount of water is equivalent to the daily water consumption of a population of close to 10,500 human beings. "Management of waste is not adequate, as some...may be dumped into rivers" (Syed et al., 2010).

4.3 Child Labor

Child labor is a "harsh reality," exposing children to risks at an age when the focus should be on education instead. The greatest number of child workers is located in India. As many as 25,000 children may be employed by the leather shoe sector in India (Syed et al., 2010).

Discussion

Leather industry portrays yet another example of governmental and corporate motivation to generate profit via the creation of problems, no matter the cost to the environment and to human beings. The murder of animals leads to the acquisition of profit from leather goods. Environmental degradation associated with leather industry contributes to human disease and additional profit. Child labor also results in profit. As elaborated in "Noun=Verb," a different book written by **Sosé Gjelaj and Elitsa Teneva**, children who are deprived from an environment rich in positive experiences suffer devastating physiological and psychological consequences later on in life. Such hardship only benefits industrial and governmental systems such as the health care and the educational sectors.

Last, but not least, according to Sosé, the awareness of the detrimental effects of leather industry on human and environmental health and wellness would suffice to influence leather goods consumer behavior in a favorable direction (i.e., choosing alternatives to leather goods) thus solving already created problems.

References

Syed, M., Saleem, T., Rehman, S. U., Iqbal, M. A., Javed, F., Khan, M. B. S. & Sadiq, K. (2010). Effects of leather industry on health and recommendations for improving the situation in Pakistan. *Archives of Environmental & Occupational Health, 65*(3), 163-172.

Chapter 5

CORPORAL VIOLATIONS

Overview

"Corporal Violations" introduces pharmaceutical drug use in livestock, aquaculture, genetically modified (GM) crops and human beings. Human health and environmental hazards pertaining to pharmaceutical use are described. The chapter concludes by distinguishing between infectious and chronic disease relative to the pharmaceutical industry.

***

Pharmaceuticals represent medical compounds used for diagnosis, treatment, mitigation or prevention of disease in animal and human populations. The global sale of pharmaceuticals in 2008 amounted to U.S. $602 billion. This figure is growing by 5% to 7% annually (Corcoran et al., 2010).

5.1 Pharmaceutical Drug Use in Livestock

The administration of antibiotics in the feed of food-producing animals is designed to promote the growth of livestock while preventing infection due to the "extremely crowded conditions that food animals are raised in" (Duckenfield, 2013). Studies demonstrate the effectiveness of antibiotic treatment in reducing livestock morbidity rate (Webb & Erasmus, 2013).

5.1.1 Estrogen

Estrogenic pharmaceuticals are sources of endocrine disruptors in the environment (der Mude, 2011). Estrogen has an effect on growth and development, reproductive behavior, sexual differentiation and pituitary hormones. Estrogen also has an impact on reproductive disorders including breast and endometrial cancer (Trudeau et al., 2005). Compounds with estrogenic properties are released into surface and ground waters directly or via water treatment plants. 90% of estrogen released in the environment results from injecting livestock with growth hormones and from failing to treat manure rich in hormones. Untreated manure enters ground and surface waters. Waters contaminated with estrogenic compounds are converted into drinking water sources (der Mude, 2011).

5.1.2 Antibiotic Resistance

There is a possibility that the misuse of sub-therapeutic levels of antibiotics in livestock feed and water is associated with antibiotic resistant infection in human beings. Infection can result from the consumption of contaminated meat, crops fertilized with animal waste and ground water polluted with animal waste. It is also possible that patients with an infectious disease do not respond to therapeutic doses of antibiotics due to the presence of antibiotic resistant genes in their organisms. Antibiotic resistance results in prolonged hospital stays, increase in health care cost, the experience of intense pain and distress. Antibiotic resistance infections can be treated in general, resulting in $17 billion to $26 billion in health care cost per year in the United States (U.S.). Infections not prone to treatment can result in death in extreme cases (Duckenfield, 2013).

The debate about antibiotic resistance continued for four decades, culminating in the European Union (EU) banning the sub-therapeutic use of antibiotics in food-producing animals in 2006 (Webb & Erasmus, 2013). Europe has banned such practice without a significant impact on the financial aspect of the industry (Duckenfield, 2013). South Africa also banned the sub-therapeutic use of antibiotics in livestock. The U.S. established a partial ban on the use of antimicrobial drugs in animals in 2012 (Webb & Erasmus, 2013). "The U.S. regulatory authorities have been slow to act" due to conflict of interests amongst key stakeholders. Discontinuing the use of sub-therapeutic antibiotics in food-producing animals will impact pharmaceutical companies negatively, yet is a "must do" given the available evidence (Duckenfield, 2013).

5.2 Pharmaceutical Drug Use in Aquaculture

Chemicals used to treat fish "are probably in many regions influencing the environment, the indigenous species, and their habitats" (Bergqvist & Gunnarsson, 2013). The primary concern of "excessive" antibiotic use in farmed fish "is that overtime it promotes the spread of resistance in both human and fish pathogens." "The full extent of antibiotic use in the industry is unclear" (Naylor & Burke, 2005).

5.3 Pharmaceutical Drug Use in Genetically Modified (GM) Crops

Some commercially released genetically modified (GM) crops contain antibiotic-resistant genes which (Ma et al., 2011) could contribute to resistance to antibiotics (Bongyu et al., 2009).

5.4 Pharmaceutical Drug Use in Human Beings

The growing consumption of pharmaceuticals on a global level contributes to their increasing discharge in surface and ground water. U.S. $1 billion of prescription drugs are discarded in the U.S. on an annual basis (Corcoran et al., 2010).

About 100 human pharmaceuticals at low levels have been identified globally in wastewater systems, groundwater, seawater and some drinking waters (Humphreys et al., 2008). Chemical pollutants such as disinfectants, sunscreens and nutritional supplements have been identified in surface waters in Europe while the evidence pointing to the presence of medications and supplements remaining in drinking water and water systems continues to grow. According to research, up to 95% of antibiotic compounds are discharged into sewage systems unaltered thus increasing the resistance of bacterial pathogens to antibiotics. Excessive concentrations of antibiotics could "affect entire food chains" (Kreisberg, 2007).

A number of pharmaceuticals leading to adverse aquatic wildlife effects have been detected in aquatic environments. Nonsteroidal anti-inflammatory drugs (NSAIDs), selective serotonin reuptake inhibitor (SSRI) antidepressants, azole antifungal drugs, lipid regulators, beta blockers, chemotherapy drugs and 17a-ethinylestradiol (EE2), an estrogen-based pharmaceutical, have been found to affect the reproductive, physiological, endocrine and immune system in fish (Corcoran et al., 2010).

Fish growth is compromised as a result of chemical pollution in aquatic environments. Impaired fish growth impedes physiological growth, olfaction, cognitive and behavioral function, locomotion which increases vulnerability to predation, affects the ability to acquire food, reproduce and survive (Groh et al., 2015).

Expensive water treatment technologies available only in certain parts of the world are capable of removing "many" pollutants from wastewater systems (Humphreys et al., 2008).

5.4.1 Infectious & Chronic Disease

Mortality rate resulting from infectious disease has decreased substantially in the last century. By 1997, infectious diseases were accountable for only 4.2% of disability-adjusted life years while chronic diseases such as cancer and coronary disease were responsible for 81% of disability-adjusted life years. Pharmaceutical companies re-directed their focus on the development of cures for chronic disease. "These chronic diseases require long term treatments and commiserate cash flows to recoup the research investments whereas antibiotics are used for shorter periods of time and do not generate the same magnitude of profits" (Duckenfield, 2013).

"How can something be so apparent and, at the same time, so disguised?"

By Sosé Gjelaj

Discussion

"Corporal Violations" depicts an ideal example of how problems are created to acquire profit and superficial solutions are generated to mask the root cause of the problem.

Quick-fix, technological fixes or miracle solutions, as indicated in the "GMO Invasion" and "Salvation" chapters, do not contribute to lasting positive outcomes. On the contrary, they delay progress. As "Paradoxical Formation" and "Salvation" propose, the real solution to a problem is within its roots. This explains why the root cause of obesity, as portrayed in "'Non-Profit' Sector and 'Food-Aid'" is rarely addressed by research. There is no profit where there is no problem.

To generalize to "Corporal Violations," first, crowded livestock conditions are set in place (i.e., created problem) which predisposes animals to infection. Next, a superficial solution or antibiotics are added to animal feed and water to reduce infection. Antibiotics are ingested by human beings consuming meat products, creating opportunities for pharmaceutical and health care profit. Antibiotics are further released into the environment, generating opportunities for additional financial gain directly through fundraising for polluted waters or indirectly through human health concerns resulting from exposure to antibiotics released in the environment. Antibiotics used as growth hormones further ensure more timely and abundant meat production, adding to acquired profit. The incentive to generate profit from crowded livestock conditions and subsequent livestock antibiotic use is specifically exemplified by the decision of the U.S. to ban antibiotics only partially despite evidence pointing to the harmful effects associated with livestock antibiotic use and the availability of an effective and lasting solution. Expanding farming space in this particular example, not adding antibiotics to feed and water, is the effective and lasting solution to the artificially-created problem, yet resolving the root cause of the problem would automatically translates into reduced opportunities for profit.

Furthermore, as we may recall from the "A Sample of Inorganic Elements" chapter, PCBs appeared to be replaced by the equally harmful PBDEs, guaranteeing the flow of profit from resulting environmental deterioration and human illness. One toxin was simply replaced by another. Similarly, in "Corporal Violations," infectious diseases appear to be substituted by chronic diseases. The form of the problem may change, but the problem always remains which ensures the continual acquisition of profit. U.S. $602 billion in global sale of pharmaceuticals in 2008 alone is an appealing incentive for the creation of problems directly linked to pharmaceuticals-related profit. All that is required is an adequately manufactured deceit in order to poison humanity and a reasonably efficient water treatment system available to those designing and implementing the profit-generating problems.

Last, but not least, an association is being established between exposure to industrial and pharmaceutical toxic substances and reproductive system impairments in both human beings and fish. One reason is that profit is being generated from environmental pollution and the development of reproductive disease. A second implication of why reproductive system disorders are a desired outcome for industries and governments would be elaborated on as part of the conclusive remarks of "Sovereign Terra."

References

Bergqvist, J. & Gunnarsson, S. (2013). Finfish aquaculture: Animal welfare, the environment, and ethical implications. *Journal of Agricultural & Environmental Ethics, 26,* 75-99.

Bongyu, M., Billingsley, G., Younis, M. & Nwagwu, E. (2009). Genetically modified foods and public health debate: Designing programs to mitigate risks. *Public Administration & Management, 13*(3), 191-217.

Corcoran, J., Winter, M. J. & Tyler, C. R. (2010). Pharmaceuticals in the aquatic environment: A critical review of the evidence for health effects in fish. *Critical Reviews in Toxicology, 40*(4), 287-304.

der Mude, A. V. (2011). Endocrine-disrupting chemicals: Testing to protect future generations. *Environmental Affairs, 38,* 509-535.

Duckenfield, J. (2013). Antibiotic resistance due to modern agricultural practices: An ethical perspective. *Journal of Agricultural & Environmental Ethics, 26,* 333-350.

Groh, K. J., Carvalho, R. N., Chipman, J. K., Denslow, N. D., Halder, M., Murphy, C. A., Roelofs, D., Rolaki, A., Schrimer, K. & Watanabe, K. H. (2015). Development and application of the adverse outcome pathway framework for understanding and predicting chronic toxicity: II. A focus on growth impairment in fish. *Chemosphere, 120,* 778-792.

Humphreys, E. H., Janssen, S., Heil, A., Hiatt, P., Solomon, G. & Miller, M. D. (2008). Outcomes of the California ban on pharmaceutical Lindane: Clinical and ecologic impacts. *Environmental Health Perspectives, 116*(3), 297-302.

Kreisberg, J. (2007). Greener pharmacy: Proper medicine disposal protects the environment. *Integrative Medicine, 6*(4), 50-52.

Ma, B. L., Blackshaw, R. E., Roy, J. & He, T. (2011). Investigation on gene transfer from genetically modified corn (*Zea mays* L.) plants to soil bacteria. *Journal of Environmental*

Science and Health, Part B, 46, 590-599.

Naylor, R. & Burke, M. (2005). Acquaculture and ocean resources: Raising tigers of the sea. *Annual Review of Environment & Resources, 30,* 185-218.

Trudeau, V. L., Turque, N., Le Mevel, S., Alliot, C., Gallant, N., Coen, L., Pakdel, F. & Demeneix, B. (2005). Assessment of estrogenic endocrine-disrupting chemical actions in the brain using *in vivo* somatic gene transfer. *Environmental Health Perspectives, 113*(3), 329-334.

Webb, E. C. & Erasmus, L. J. (2013). The effect of production system and management practices on the quality of meat products from ruminant livestock. *South African Journal of Animal Science, 43*(3), 413-423.

Chapter 6

TEMPLATE FOR EXTINCTION

Overview

"Template for Extinction" discusses water contamination, whether it be drinking water, surface or groundwater. Common contaminants such as fluoride, manganese and uranium are explored. Water treatment and distribution processes are further elaborated upon. Iodine is proposed as a viable water disinfectant. Freshwater resources and pertinent to freshwater processes and procedures are further explored. "Template for Extinction" ends by presenting on the topics of water access and privatization.

***

There is a general consensus that the "water problem" is not the result of water scarcity in the hydrosphere, but is rather due to the ever-increasing gap between the rate of used water self-renewal process and the present and future rate of water consumption. The consequence is soil, surface and ground-water contamination with anthropogenic and natural chemicals (Zoller, 2006). 1.1 billion people are without access to safe water in the world (Dagdeviren, 2008). Millions of people die on an annual basis from exposure to contaminated drinking water (Huby & Stevenson, 2003).

6.1 Water Health Benefits

"Water is a vital resource for life and access to safe drinking water is a basic right of every individual" (Mulamattathil et al., 2014). "Water represents a critical nutrient, the absence of which will be lethal within days." 55% of the body weight of the elderly and 75% of the weight of infants consists of water.

Water is obtained directly as a beverage, from food and to a small degree, from the oxidation of micro-nutrients. Water is associated with cognition. Mild degree of dehydration (lack of water) could contribute to impairments in mood and cognitive function which are reversed once water is reintroduced into the body. One of the risk factors of delirium is dehydration. Dehydration could also lead to headache (Popkin et al., 2010). Consumption of water is linked to beneficial health outcomes including obesity prevention, decrease in energy intake and improved childhood cognitive function (Patel et al., 2014).

6.2 Water Contamination

"Water-borne disease continues to pose a major threat to public health both in the developed and developing world". According to a 2002 World Health Report, unsafe water, hygiene and sanitation contribute to the death of approximately 1.7 million people each year, primarily in developing countries as a result of infectious diarrhoea. "Nine out of ten such deaths are in children." According to 2002 estimates by the World Health Organization (WHO), approximately 2.6 billion people or 42% of the population in the world lack adequate sanitation while 17% lack improved water resources (Hamner et al., 2006).

6.2.1 Groundwater

Approximately one third of humanity in the world acquires drinking water from groundwater (Cheng et al., 2013). Groundwater quality in the Mediterranean area is deteriorating in regions with great pressure on aquifers. Pressures include depletion, wastewater discharge, release of pesticides, mine tailings contamination and salinisation among others (Garcia-Prieto et al., 2012).

One study revealed that only 6.7% of groundwater in Zhengzhou City, China, was considered permissible for consumption and the remaining was viewed as desirable. According to a 2010 China Geological Survey (CGS), 90% of groundwater in China is contaminated and 60% is "seriously polluted". A different study demonstrated that 34% of groundwater in Thanjavur City, India, was not considered as suitable for ingestion (Cheng et al., 2013).

About 104 million Americans consume groundwater. 61% of water-borne disease outbreaks in 2007 and 2008 in the United States (U.S.) was attributed to groundwater consumption. Small municipal groundwater systems supply non-treated drinking water to 20 million Americans. There is a growing awareness that groundwater can be contaminated by human viruses (Uejio et al., 2014).

6.2.2 Farm Ponds

Farm ponds are threatened on a global level as a result of eutrophication, chemical pollution, changes to land use, invasion by non-native species and physical destruction. Pond draining is a common pond management method believed to remove invasive fish and improve water quality. It is shown that pond draining has "little effect on invasive fish control, water quality, or aquatic biodiversity" (Usio et al., 2013).

6.2.3 Surface Waters

Sodium chloride (NaCl) contained in road salts "represents the largest chemical loading to the Canadian surface waters." 6.8 million tons of road salts were sold for highway de-icing in 2003. Elevated levels of chloride could contribute to surface water quality degradation as once dissolved, the salts enter drainage systems, soils or water streams. Increasing chloride levels in surface waters within the Greater Toronto Area (GTA) in Canada were demonstrated as part of a study. The researchers stated that "gathering salt usage data from the municipality officials proved very challenging and it was an insurmountable obstacle because the municipalities were the sole proprietors of public salt usage data" (Amirsalari et al., 2013).

A great number of chemical pollutants are contained in pharmaceuticals and personal care products (PPCPs) which "continue to grow worldwide." Wet weather could contribute to the leaking of PPCPs from garbage disposal landfills into the environment such as in surface water systems. A recent study demonstrated the presence of more than 100 kinds of PPCPs, including estrogen compounds, in "significant concentrations" in water systems. The rate of removal of PPCPs from the environment is "overwhelmed" by their rate of replacement. "The increasing amount of PPCPs in the environment is a source of great concern" (Kreisberg, 2007).

6.2.4 Rivers

Rivers are the main conduits of water, dissolved material and particulates from the surface of the earth to the ocean (Hedley et al., 2010). The majority of rivers in the world are shared by two or more countries. Industrial and population growth have "challenged the quality of these rivers." Negotiations concerning trans-boundary pollution of international rivers could be difficult and time-consuming (Dieperink, 2011).

6.2.4.1 Rhine River

The 1,326 km long Rhine river flows through Germany, France, the Netherlands and Switzerland, transporting waste water (containing chloride) from a region with more than 58 million people and an "impressive number of industries" (Dieperink, 2011).

6.2.4.2 Ganges River, Varanasi, India

An estimated 200 million liters or more of untreated human

sewage is discharged on a daily basis into the Ganges River in Varanasi, India. A significant association was established between water-borne/enteric disease and the use of Ganges River for brushing teeth, washing eating utensils, laundry and bathing. In 1986, the Indian government initiated the Ganga Action Plan (GAP), designed to manage sewage contamination of the Ganges River. Phase I from the project was completed in 1993. Despite the initiative, Ganges River continued to be highly polluted with untreated human sewage. A recent governmental report states that 200 million liters or more of untreated sewage presently enter the Ganges River each day. The Government of India has committed to initiate a second phase of GAP to deal with Ganges River pollution (Hamner et al., 2006).

6.2.5 Tourism

1 in 8.1 Americans are employed in the tourist industry. The U.S. tourist industry contributes approximately $200 billion to U.S. exports, which is greater than the export value of all U.S. computers, agricultural products, aircraft and telecommunication equipment. Beaches represent the leading destination for tourists. 9% of Wales residents are employed in the tourist industry, which is worth 3 billion pounds per year. The aesthetic appearance of beach water and surroundings shapes tourists' initial perception of the coastal environment (Tudor & Williams, 2008).

Aquatic litter (i.e., plastic; paper; aluminum) is perceived as a worldwide environmental challenge (Kordella et al., 2013). Tudor & Williams (2008) investigated the perception of 1824 beach users at 20 United Kingdom (UK) beaches regarding coastal pollution. 589 participants ranked sewage-related debris (SRD) as the most offensive form of beach pollution. Beach and water oil pollution was ranked second. These were followed by foam/scum, debris, smell, litter and discolored water (Tudor & Williams, 2008). Kordella et al., (2013) demonstrated that water navigation was the primary reason for litter pollution on 80 Greek beach areas.

6.2.6 Restaurant Industry

Waste generation, improper package and product disposal, the use of chemicals and astounding amounts of water are environmental repercussions from the restaurant industry. High consumption of water occurs during inappropriate food processing techniques such as thawing food products under running water which not only poses environmental risks, but also human health risks due to sanitary-hygienic considerations. The WHO recommends thawing food only in refrigerators or cool areas instead (Martinelli et al., 2012).

Wastewater from restaurants and food processing establishments contributes to substantial environmental contamination. Electrochemical technologies including electroflotation (EF) and electrocoagulation (EC) are cost-effective, efficient and environmentally compatible in removing a number of pollutants from wastewater (Qin et al., 2013).

6.3 Water Contaminants

6.3.1 Fluoride

Fluoride is the 13th most common element present on the earth's crust (Al-Saleh & Al-Doush, 2000). Fluoride is an air, soil (Kamaluddin & Zwiazek, 2003) and water (Yakub & Soboyejo, 2013) pollutant (Kamaluddin & Zwiazek, 2003) that enters the environment via anthropogenic activities. Industrial operations represent the major anthropogenic source of groundwater fluoride in the past 60 years. The global fluoride levels in groundwater range between 1 and 35 mg/L (Inyang, 2004).

6.3.1.1 Health Effects

In the past 15 years, the addition of fluoride to drinking water as a means to prevent dental cavities has "generated scientific controversies and policy disputes" (Inyang, 2004). Excessive fluoride levels in drinking water threaten the health of hundreds of thousands of people in the world. Exposure to fluoride from drinking water above the international acceptable standard of 3.5 mg/L causes fluorisis (Bagh et al., 2006), characterized by pitting, mottling and staining of the dental enamel (Yakub & Soboyejo, 2013). Dental fluorosis is not perceived as a health risk (Inyang, 2004). Besides drinking water, other sources of fluoride that may add to fluoride intake and thereafter the development of fluorosis include fluoride supplements, fluoride toothpaste, infant formula, beverages containing fluoridated water, food produce grown in fluoridated soil or irrigated with fluoridated water and cow' milk from cows raised with fluoridated water and feed (Erdal & Buchanan, 2005).

Fluorosis subsequently leads to gastrointestinal ailments and deformity of organs (Bagh et al., 2006). The Environmental Protection Agency (EPA) has established a maximum contaminant level (MCL) of fluoride in drinking water at 4.0 mg/L (Inyang, 2004). "The ever-increasing fluoride levels in water, food and air pose a great threat to human health and to the environment" (Gupta et al., 2007). Exposure to fluoride levels between 10 mg/L and 20 mg/L may contribute to the development of an "incurable" disease, skeletal fluorosis (Bagh et al., 2006). Exposure to excessive fluoride levels may cause renal toxicity and epithelial lung cell toxicity and reproductive defects (Sharma et al., 2008). Elevated levels of fluoride in drinking water is linked to a decrease of birth rate. Infertility was observed in the regions of India where fluorosis is endemic (Gupta et al., 2007).

6.3.1.1.1 United States (U.S.)

Fluoridation of drinking water in the U.S. was initiated in 1945 (Christoffel, 1985). A 1974 survey revealed that 26% of the U.S. dentists who took part in the study prescribed fluoride tablets while 17% prescribed both fluoride supplements and vitamins. A 1982 survey demonstrated that 60% of dentists surveyed prescribed fluoride tablets (Levy et al., 1984). Levy et al. (1984) investigated whether North Carolina health care providers, including dentists, prescribed fluoride supplements after sampling drinking water for fluoride content. The analysis was conducted between 1982 and 1983. Findings from the study revealed that a small percentage of health care providers were assaying drinking water for fluoride content. Therefore, many children who required fluoride supplementation were not receiving it while another group may had been receiving fluoride tablets without proper sampling of the drinking water for fluoride content. The identified patterns were pertinent to other regions of the U.S. (Levy et al., 1984).

The rate of dental fluorosis in the U.S. has increased in the past 30 years. Childhood dental fluorosis may develop upon exposure to fluoride during the dental enamel formation period. A survey conducted in the U.S. between 1986 and 1987 revealed that 22% of surveyed children had developed fluorosis. A different study demonstrated that in 1998, 69% of children between the ages of 7 and 11 had fluorosis. 78% of children in fluoridated communities in the U.S. suffered from fluorosis while the prevalence of childhood fluorosis in non-fluoridated communities ranged between 3% and 45% (Erdal & Buchanan, 2005). Fluoride further affects children's growth by accumulating in bones and reducing calcium uptake (Wang et al., 2007).

A number of "reputable" medical and scientific organizations in the U.S., including the American Dental Association, the National Academy of Sciences, the American Medical Association and the American Pharmaceutical

58

Association "have endorsed the addition of fluoride to drinking water" (Inyang, 2004). 162.1 million Americans (over half of the U.S. population) were consuming fluoridated drinking water in 2000 (Macek et al., 2006). Testing of public drinking water for fluoride concentrations is required in the U.S. Municipalities are mandated to treat drinking water to comply with MCL fluoride recommendations. Despite such arrangements set in place in relation to fluoride in drinking water in the U.S., "some civil libertarians view fluoridation as an enforced and unconstitutional medication of the American public."

Opponents of drinking water fluoridation propose that fluoridated water could cause fluoride poisoning, sterility, birth defects, allergic reactions, genetic damage and/or cancer. In 1982, a trial judge in Illinois ruled that the Illinois law authorizing the fluoridation of public drinking water "was an unreasonable exercise of the police power and therefore unconstitutional and invalid." In response, the Illinois Supreme Court reversed such ruling, finding that "plaintiffs have shown, not that the risk was so great that fluoridation was unreasonable, but that the question was shown to be debatable. Under these circumstances plaintiffs have failed to show an unreasonable exercise of the police power." The Supreme Court subsequently ruled the continuation of fluoridation in the state of Illinois (Christoffel, 1985).

6.3.1.1.2 China

Fluoride is a widespread contaminant of water supplies in China (Cheng et al., 2013).

6.3.1.2 Solution

Clay-hydroxyapatite (C-HA) is a new technique for removing fluoride from drinking water. It is designed to filter out not only fluoride particles from contaminated drinking water, but also microbial pathogens, such as E. coli (Yakub & Soboyejo, 2013). Further, Makhado et al. (2006) examined the ability of wood ash to decrease fluoride levels in drinking water. Findings from the study revealed that small amounts of ash from trees were effective in reducing fluoride content from drinking water.

6.3.2 Arsenic (As)

Contamination of groundwater with As is prevalent "in almost all countries" (Chakraborty & De, 2009). 200 million people in the world are exposed to As levels in drinking water exceeding the 10 mg/L recommendation for drinking water provided by the WHO. 45 million people are exposed to extremely high levels of As in drinking water (George et al., 2014). It is possible that more than over 100 million people worldwide are exposed to iAs at concentrations exceeding 10 mg/L which is the standard for drinking water in a large number of countries. Drinking water from contaminated underground wells is the primary route of iAs exposure (Benbrahim-Tallaa & Waalkes, 2008). A large number of the U.S. and worldwide population acquires its drinking water from private and unregulated wells (Davey et al., 2008).

6.3.2.1 Bangladesh

The 2001 British Geological Survey estimated that approximately 50 million people in Bangladesh consumed drinking water from tube wells which exceeded the WHO recommended As levels of 10 mg/L (Li et al., 2008).

6.3.2.2 Peru

George et al. (2014) investigated As contamination of 151 surface water and groundwater sources in 12 districts of Peru. Peru is one of the major producers of arsenic in the world, used mainly in the production of insecticides and pesticides. Peru is also a major producer of copper, silver and gold. The national regulatory standards for As in drinking water in Peru are based on WHO recommendations (i.e., 10 mg/L). The As concentration exceeded 10 mg/L in 86% of the samples. As levels exceeded the Bangladesh guideline of 50 mg/L in 56% of the samples. Drinking water in sampled districts in Peru is therefore widely contaminated with As, posing public health risk. The Ministry of Health in Peru is charged with ensuring national drinking water quality. Yet, "no systematic attempts have been made to conduct countrywide arsenic surveillance or mitigation." The results from the present study revealed "an alarming public health threat that needs to be addressed immediately."

6.3.2.3 U.S.

About 98% of the U.S. population consumes drinking water with less than 10 mg As/L. About 2% of the U.S. population was projected to have As levels higher than the revised MCL in 2006. Due to the association between inorganic arsenic (iAs) and cancer, the U.S. EPA lowered the maximum contaminant level (MCL) of As in water from 50 to 10 mg As/L (Hughes, 2006).

6.3.2.4 Health Effects

Exposure to As concentrations in drinking water greater than 50 mg/L could lead to increased risk of lung cancer, cardiovascular, neurological, respiratory disease, skin lesions and increased rates of mortality (George et al., 2014). Millions of people in the world suffer from cancer and other diseases linked to exposure to water contaminated with iAs (Hernandez-Zavala et al., 2008). Millions of human beings globally consume drinking water with iAs contamination levels over 100 mg/L (Hughes, 2006).

Chronic As intake from drinking water has been strongly correlated with increased prevalence of a number of cancers, diabetes, developmental and reproductive health concerns. As is further a powerful endocrine disruptor even at low concentrations (Davey et al., 2008). Chronic consumption of As-contaminated water contributes to the development of vascular disease, including cardiovascular and peripheral disease. Study findings revealed that replacing As-contaminated water with safe drinking water could reverse the resulting from arsenosis peripheral vascular disease (Pi et al., 2005).

Children's growth, development and intellectual function are influenced by a number of factors including heredity, education, nutrition, geography and society. Arsenic exposure from drinking water has been linked to reduced intelligence in children. A study found out that exposure to 5 mg/L of As in drinking water contributed to reduced intellectual function in a 10-year old Bangladesh child (Wang et al., 2007). Studies have demonstrated a relationship between consumption of As in drinking water below 200 mg/L, cognitive function, reproductive effects and cardiovascular disease (Schuhmacher-Wolz et al., 2009).

Davey et al. (2008) investigated whether As could disrupt gene regulation via the thyroid hormone (TH) and/or the retionic acid (RA) receptors. Both TH and RA are vital in normal development and function while their impairment is linked to a number of disease processes. Findings revealed that As exerts a significant effect on TR and RAR gene regulation at very low concentrations. The results from the study suggested that exposing young people to As during critical developmental periods could lead to "significant long-term consequences."

6.3.3 Manganese (Mn)

Manganese is an essential nutrient in both animals and

humans (Wasserman et al., 2006) in recommended doses (Hafeman et al., 2007). "The 1958 WHO *International Standards for Drinking-water* suggested that concentrations of manganese greater than 0.5 mg/L [500 μg/L] would markedly impair the potability of the water."

In 1993, the WHO established a 500 μg/L standard for Mn in drinking water. In 2004, the 1993 drinking-water guideline for Mn was reduced to 400 μg/L. The WHO removed the 400 μg/L Mn standard for drinking water in 2011, attesting that "this health-based value [400 μg/L] is well above concentrations of manganese normally found in drinking-water, [so] it is not considered necessary to derive a formal guideline value" (Frisbie et al., 2012).

It is a known fact that exposure to manganese (Mn) via inhalation has neurotoxic effects in adults. Some children in the U.S. and Bangladesh are at risk of neurotoxic effects induced by Mn. Approximately 6% of U.S. household wells contain Mn that exceeds 300 μg Mn/L (Wasserman et al., 2006) while Bangladesh groundwater is commonly contaminated with Mn. A study revealed that 35% of Bangladesh water samples exceeded the 1993 WHO levels of 0.5 mg/L (Hafeman et al., 2007). Research findings reveal that exposure to Mn-contaminated drinking water resulted in reduced cognitive function (i.e., reduced Full-Scale, Performance, and Verbal raw scores) in 142 10-year-old children in Bangladesh (Wasserman et al., 2006).

Infant mortality rate in Bangladesh is "extremely high," amounting to 54.0 per 1,000 live births in 2000. Neonatal tetanus and acute lower infections are the most common causes of infant mortality. The association between Mn in drinking water and infant mortality was investigated. Children who were exposed to water Mn at levels exceeding 0.4 mg/L had an increased risk of mortality (Hafeman et al., 2007).

6.3.4 Uranium

Upon prolonged ingestion, uranium accumulates in bones and kidneys (i.e., uranium toxicity). When consumed through drinking water, uranium leads to the excretion of essential to the bone structure minerals, calcium and phosphate. Calcium excretion could lead to bone resorption. Uranium-rich ores and highly soluble uranium contribute to elevated uranium levels in groundwater. The established by the WHO guideline for uranium in drinking water is 15 mg/L (Kurttio et al., 2005).

6.3.5 Nitrates

Exposure to nitrates occurs via the consumption of drinking water, vegetables, medications and processed meat. A possible association between nitrate intake and thyroid dysfunction, particularly the development of childhood goiter, has been established. The present study confirmed that childhood goiter developed as a result of exposure to nitrate levels in drinking water (Gatseva & Argirova, 2005).

6.3.6 Atrazine

Atrazine, a widespread herbicide used on golf courses, lawns and crops, "is now among the most common pollutants in drinking water" (der Mude, 2011).

6.3.7 Aluminum Lactate (Al)

Chronic exposure to low levels of aluminum lactate (Al) may increase the risk of developing Parkinson's disease (PD) and other age-related neurogenerative diseases of the central nervous system (CNS). Elevated levels of Al in drinking water have been linked to a greater risk of developing Alzheimer's disease (AD). From 13 published epidemiological studies, 9 have demonstrated a statistically significant positive correlation between Al levels in drinking water and incidence of AD. Epidemiologic studies have further established a link between exposure to Al and PD. Al is able to induce oxidative stress (Li et al., 2007).

6.3.8 Water Viruses

Adenoviruses are human pathogens that contribute to the development of a number of diseases including respiratory and gastroenteritis disorders. Adenovirus infection could result from consuming contaminated water or inhaling aerolized particles as part of water activities. A great number of adenoviruses enter the raw sewage system as they are excreted in human feces. Adenoviruses could also be found in large numbers in polluted waters. A great number of adenoviruses have been identified in rivers, coastal waters, swimming pools and drinking water. Some water treatment methods are unable to eradicate adenoviruses, which may pose public health risks provided contaminated with adenoviruses water is used as drinking water (Dong et al., 2010).

Rotavirus is the main cause for the development of diarrhoeal disease in the world. It especially affects children. Almost all children at the age of 5 are impacted by rotaviruses. On an annual basis, rotaviruses contribute to the onset of 114 million diarrhoea episodes, 2.4 million hospital admissions and 600,000 deaths in children younger than 5 years of age. Rotavirus has been perceived as the causative agent to a number of waterborne outbreaks in developed countries, suggesting "the survivability of rotavirus in water" (Rutjes et al., 2009).

6.3.9 Estrogen

Sewage treatment plants (STPs) contain a great amount of contaminants that are not fully eradicated during the water treatment process. Waste water from STPs is viewed as a primary source of estrogenic water contamination. It is discharged into rivers following the waste water treatment process (Pillon et al., 2005).

6.3.10 Hardness

Findings concerning the association between the hardness of drinking water (determined by the amount of calcium and magnesium) and mortality related to stroke or ischemic heart disease (IHD) have been controversial. The present study observed no such association (Leurs et al., 2010).

6.4 Iodine

Iodine is a fundamental nutrient that ensures optimal thyroid function (Backer & Hollowell, 2000). There are approximately two billion people worldwide who are iodine deficient (Hoddinott et al., 2008). Toxic iodine levels are extremely high, ranging between 200 and 500 mg per kg (Fisch et al., 1993). Toxic reactions to excessive amounts of iodine include iodide fever, sialadenicis, nausea, diarrhea and epigastric pain. Thyroid disorder such as hypothyroidism with or without goiter is the main health consequence resulting from excessive ingestion of iodine. Most people can ingest large amounts of iodine without suffering thyroid dysfunction consequences.

There is substantial controversy regarding the maximum safe iodine dose and length of use. Neither the maximum recommended daily dose of 2 mg/day nor the maximum recommended dietary dose of 3 weeks are grounded on firm evidence. The maximum safe daily intake of iodine over a long period of time has not been determined. Studies have demonstrated that daily ingestion of iodine between 1 and 2 mg is safe for the majority of people and that even higher doses are typically well tolerated (Backer & Hollowell, 2000). The long-term iodine intake of less than 2 mg per day is considered safe (Fisch et al., 1993). Individual response to iodine (typically based on an underlying thyroid condition) rather than an iodine consumption threshold appears to be responsible for the health effects relative to iodine ingestion dose and time frame. The majority of humanity may use iodine (1-2 mg/L) for years provided their thyroid function is monitored (Backer & Hollowell, 2000).

6.4.1 Iodine Deficiency & Solution

Iodine deficiency may contribute to the development of a number of health conditions including hypothyroidism, goiter, spontaneous abortion, impaired mental function, stillbirths, congenital anomalies, increased prevalence of perinatal and infant mortality, neurological and myxedematous cretinisms (Fisch et al., 1993).

Iodine supplementation has contributed to reduced rates of goiter and hypothyroidism due to iodine deficiency in major parts of the world. Dietary sources of iodine include eggs, meat, dairy, bread, or seaweed (Backer & Hollowell, 2000).

Fisch et al. (1993) developed a collective prophylactic method against iodine deficiency. The method consists of the installation of a silicone elastomer in a well which releases iodine at the recommended iodine daily dose. The technique was tested in a village in West Africa where goiter was endemic. After 12 months of application, the prevalence of goiter had dropped from 53.2% to 29.2%. Such innovative technology can be adapted to all sources of water supply. It has the potential to "contribute to the eradication of iodine deficiency" (Fisch et al., 1993). "Once implemented, water iodization is almost invariably associated with community-wide iodine repletion" (Foo et al., 1998).

6.5 U.S. Water Treatment

U.S. drinking water regulation and treatment differ by water source and water treatment method. The Safe Drinking Water Act and Groundwater Rule require surface and ground water monitoring and treatment of surface water. Water treatment "sanitizes, filters or inactivates most pathogens." Pathogens resistant to water treatment "cause a sizeable GI burden." 2 to 19 million Americans suffer from gastrointestinal illness (GI) every year as a result of consuming contaminated drinking water. Past GI infection could lead to adverse long-term health effects (i.e., irritable bowel syndrome; reactive arthritis) that "accrue significant public health cost." 27% of U.S. public water wells are "expected" to be polluted with human viruses. Municipal water systems that do not undergo water treatment tend to contribute to higher rates of water-borne illness (Uejio et al., 2014).

Mycobacteria are widespread environmental bacteria that particularly affect people with compromised immune system. Mycobacteria may resist water treatment and disinfection and live in waters and soil. Mycobacteria may not only survive, but also grow in biofilms in pipeline water systems (Rasanen et al., 2013).

"After the terrorist attacks on the United States in 2001, concerns about the intentional contamination of drinking water systems have increased." Purposeful contamination of drinking water with anthrax is of concern to water networks as Bacillus anthracis spores could attach to biofilms and not respond to disinfection (Hosni et al., 2009).

6.5.1 N-nitrosodimethylamine (NDMA)

N-nitrosodimethylamine (NDMA) is a N-nitrosamine. N-nitrosamines are recognized as probable carcinogens. NDMA is the most commonly detected product of water treatment. The WHO has established drinking water guidelines for NDMA of 100 ng/L while the recommended level established by Health Canada (HC) is 40 ng/L. When detected in drinking water, NDMA is typically less than 10 ng/L. Approximately one third of the U.S. population or 100 million people that use public water services are exposed to detectable NDMA levels in drinking water. Beer consumption contributes 33% to dietary NDMA intake (Hrudey et al., 2013).

6.5.2 Disinfection by-products (DBPs)

The primary goal of water disinfection in the U.S. is to ensure the distribution of chemically and microbiologically safe water. Chlorine disinfection is the major disinfection method used in U.S. water distribution systems. However, chlorine could interact with natural organic matter (NOM), producing disinfection by-products (DBPs) such as chloroform (Clark & Sivaganesan, 2002) trihalomethanes which have carcinogenic properties (Michael et al., 1981). Other forms of DBPs have been identified in the U.S. since 1974, raising concerns over the potential public health risks that could result from their exposure (Clark & Sivaganesan, 2002). Concerns have been raised regarding possible adverse reproductive effects such as low birth weight, stillbirth, congenital anomalies and spontaneous abortion. Findings from studies have been, however, inconsistent. An inverse significant association has been found between trihlomethan exposure and neural tube, urinary tract, respiratory and cardiac defects. Yet, other studies have failed to demonstrate such correlation (Nieuwenhuijsen et al., 2008).

6.5.3 Iodine Disinfection

"Iodine has been used to ensure the safety of potable water since the 1940s, when the military developed a tablet formulation for use by troops in the field. Widespread use followed in the civilian population. Because of the ill-defined risk of iodine affecting thyroid function and because other means of water treatment are often available, the World Health

Organization (WHO) and the U.S. Environmental Protection Agency (EPA) recommend that iodine be used for short-term or emergency use only for water treatment. However, many people use iodine for much longer periods because of the convenience and effectiveness of the products...there are remarkably few reports of resulting clinical thyroid disorders" (Backer & Hollowell, 2000).

6.6 U.S. Water Distribution

The formation of biofilms on pipe walls in drinking water systems takes place despite the use of disinfectants. Their presence could contribute to the development of bacteria and microorganisms and pose public health risks (Hosni et al., 2009).

6.7 Fresh Water

"The world is running out of fresh water suitable for human use." About 75% of Earth's surface is occupied by water. 3% of global water cover is fresh water. Of the 3% fresh water, 69% is contained within glaciers and ice caps, 30% is located in soil and groundwater and some fresh water is found on surface water systems such as lakes and rivers (Yuknis, 2011).

There has been a growing unsettling regarding the decline of usable water resources due to an increase in usable water consumption by a rising population, irrigation and industries. "Currently, the world is moving towards a water crisis." Africa, the Middle East and Southern Asia are the areas in the world most severely affected by usable water shortage (Feleafel & Mirdad, 2013). Water resources in Hong Kong are "inadequate" for a population of 6.7 million people. Therefore, a great volume of fresh water is imported from Guangdong Province in China (Chen & Wong, 2004).

Based on estimates by the Department of the Interior, 18% of U.S. fresh water is used by domestic households, 59% -

by industries and 33% is allocated to the agricultural sector (i.e., irrigation). California cultivates 50% of fresh produce in the U.S. Increasing agricultural demands resulting from a growing population has "strained California's water resources," making the state of California especially vulnerable during times of droughts (Yuknis, 2011). A California survey revealed that 75% of households depended on bottled or filtered water as their major source of drinking water (Davis, 2005).

6.7.1 Lake Okeechobee, Florida, U.S.

The South Florida Water Management District also referred to as the Water District, is a regional governmental agency "responsible for managing and protecting the water resources of South Florida."

Lake Okeechobee is situated in southern Florida, in proximity to the Everglades Agricultural Area. The Everglades Agricultural Area contains canals built by the U.S. Army Corps of Engineers to collect runoff and rainwater from proximal residential and industrial areas. Collected water contains chemical pollutants. The Water District is responsible for managing three pumping stations that "pump water from the lower levels of the canals into the higher waters of Lake Okeechobee...the contaminants already in the water from the canals are not removed before the water is deposited in the lake." The freshwater originally contained within Lake Okeechobee is currently mixed with pollutants from the contaminated water pumped into it from the U.S. Army Corps canals. The Water District is presently collaborating with the EPA to improve the water quality in the Everglades National Park. Fresh water in the Everglades National Park has flowed from Lake Okeechobee to Florida Bay each year during the wet season for centuries (Cabrera, 2014).

6.7.2 Bay-Delta Estuary

71

Estuaries transport biomass and energy from fresh water systems to marine domains (Shou et al., 2013). The Bay-Delta is the largest estuary in Western U.S. and the largest source of water in California. The Bay-Delta is a habitat to over 750 plant and animal species. The estuary's ecosystem health (i.e., increase in invasive species), water quality and quantity have been gradually declining for more than a decade. Such continual deterioration has spurred disputes between environmentalists, agriculturalists and urban water consumers over how to address these concerns. California is a home to more than 36 million people and to the largest agricultural production in the U.S., generating more than $36 billion per year. A three-year drought in California characterized by below-average rainfall contributed to a water crisis as the increasing demand for water for a growing population and large agricultural production systems did not meet the water supply. More than 60% of the 5.9 million acre-feet of water supplied by Bay-Delta each year is allocated to agricultural production in California. The growth of the agricultural production requires greater water supply from Bay-Delta, straining its water supply and contributing to its decline. The Bay-Delta estuary supplies drinking water to two-thirds of California households, the equivalent of 24 million people. While the Bay-Delta water supply is not increasing, the population in California is growing, making it challenging to meet the rising demand for water supply (Wimberger, 2009).

6.7.3 Bottled Water

As freshwater is being "stretched" to meet agricultural, industrial and growing human population demands, "the shortage of safe and accessible drinking water will become a major challenge in many parts of the world." Bottled drinking water, available in both developing and developed nations, is promoted as a solution to concerns related to safety and quality of drinking water. Healthy drinking water is characterized by the removal of harmful chemicals from water while preserving the presence of essential minerals.

17 brands of bottled drinking water in India were examined to determine compatibility with drinking water recommendations established by the WHO and the EPA. Findings revealed that the greater part of bottled drinking water was over-treated and deficient in adequate essential mineral concentrations. Essential minerals such as calcium and magnesium have been associated with decreased risk of osteoporosis and sudden death respectively while both have been suggested as being protective factors against gastric cancer. Lead levels were greater than recommended in 7 bottled drinking water samples, potentially posing risks to public health. Long-term consumption of mineral-deficient water brands would necessitate the additional intake of mineral supplements to account for the need of the human body of minerals. Minerals in drinking water are absorbed more efficiently by the organism compared to mineral intake from food (Mahajan et al., 2006).

6.8 Water Access & Privatization

6.8.1 Water Access

In a document released in the year 2000, the United Nations Children's Fund (UNICEF) and the WHO stated that 82% or 4.9 billion people in the world had access to safe water, 60% or 3.7 billion people had access to sanitation facilities, 94% of the urban and 71% of the rural population had access to safe water, 86% of the urban and 38% of the rural population had access to sanitation facilities. Access to safe water was defined as the availability of 20 liters of clean water per individual within a distance of 1 km (Chenoweth & Bird, 2003).

6.8.2 Water Privatization

Since the 1980s, privatization of public services (i.e., transportation; telecommunications) has spread globally (Davis, 2005). "In recent years, a significant trend has been the transfer of water management from public to private operation" (Morris & Cabera, 2003). "The neoliberal solution to problems in the water sector has been privatization," considered to be a "poor policy prescription" (Dagdeviren, 2008). Privatization of the water and sanitation (W&S) sectors has been controversial. By the end of 2000, at least 93 nations have had at least part of their water and wastewater sectors privatized. Private water companies served 51 million people in 1990 and 300 million people in 2002 (Davis, 2005). The privatization of water services has become the "de facto policy" in a number of countries, particularly in developing countries. One such nation is Zambia (Dagdeviren, 2008).

6.8.2.1 Zambia

As of 2006, there were 10 commercial water and sanitation companies (WSC) in Zambia, serving 90% of the urban population. In 1990, prior to the privatization of the water sector, 73% of the population in Zambia had access to water. This number dropped to 53% in 2005, causing a large number of households to depend on boreholes, wells and public taps for water supply. Commercialization of water services in Zambia not only did not contribute to increased access to water, but on the contrary, since 2001 it led to more than 20% average reduction in access rate. In 1987, the government of Zambia had invested U.S. $16.6 million in the water sector which was 30 times more than the amount spent in 2002. The government spent 2% to 12% of its budget on water services between 1998 and 2002. Governmental expenditure on urban water services in Zambia between 1994 and 2003 was not geared towards ensuring safe water access for a greater proportion of the population, but was rather directed towards maintenance of the commercialized water system infrastructure. All utilities in the urban regions of Zambia currently charge higher tariffs to improve their operations. A large number of the urban population in Zambia is not able to afford rising utilities tariffs. The privatization of water services in Zambia "seems destined to be a failure without up-front investment." Taking into consideration the substantial budget required for water service maintenance and the limited funding of the governments of developing countries, "development assistance by donor countries has to be an important source of funding" (Dagdeviren, 2008).

6.8.2.2 Rural Areas

13 million children younger than 5 years of age die from inadequate sanitation and other causes linked to poverty on an annual basis. Contaminated water and air are the two main causes of childhood diarrhoea and respiratory illness, "the two biggest killers of poor children." Rural areas experience some of the greatest challenges in relation to water quality. Resources in rural regions of the world are "not readily available for projects to improve availability, consumption rates and water quality," leading to developing nations welcoming external agencies' interventions. A large number of wealthy governments and donor agencies have invested substantial resources in improving water supplies in financially disadvantaged rural areas. Yet, a great number of such projects fail to achieve that objective. Financial and economic constraints, inadequate technology, lack of scientific or technical information are all proposed reasons for such failure. Findings from rural water supply studies demonstrate that the "dig-install-depart" intervention model "often leaves communities worse off than they would otherwise have been. Their dependence has increased but not their supply" (Huby & Stevenson, 2003).

6.8.2.3 China

In Xinjiang, China, the idea of water being treated as an economic good "has not contributed to a fairer and more efficient use of scarce water resources, but rather it has been applied to achieve other political and economic goals as well as to strengthen a powerful and rapidly growing bureaucracy." Farmers in Xinjiang question why government water management officials require them to pay very high water fees when the quality of water supply that they receive is "increasingly deteriorating." Inadequate water management contributes to low crop production, which is the major cause of poverty for a number of farmers. According to Bazhou Water Management Department (BWMD), water fees collected by governmental water management organizations accounted to 2.8 million yuan in 1990 and 52 million yuan in 2002. Water fees are used on activities such as water infrastructure, staff training, salaries, repairs, equipment, prevention of flooding. However, there are "no clear written regulations" regarding how much money is to be spent on a particular activity. "Ultimately, water reforms may easily end up as yet another way of taxing China's farmers in order to build up a large and affluent water bureaucracy" (Yuling & Lein, 2010).

6.8.2.4 Mexico

The privatization process for water servicing began in the early 1990s in Latin America. A number of factors have influenced the water privatization process in Mexico. By the 1980s, the provision of water services in Mexico has emerged as a "major social problem." In 1990, about 16.7 million people in Mexico did not have access to drinking water partially as a result of institutional, financial and technical issues pertinent to the water servicing sector (Morris & Cabera, 2003).

6.8.2.5 U.S.

14% of the U.S. population receives W&S services from private companies. Such percentage has remained relatively unchanged since the 1940s (Davis, 2005).

6.8.2.6 Research Findings

Evidence from empirical research in the area of water and sanitation sector privatization "is largely inconclusive." One such reason is the difficulty of establishing a control group. One review of 20 studies comparing public and private water and sanitation services in the U.S., England, Wales and France between 1977 and 2000 revealed "no compelling evidence to date of private utilities outperforming public utilities or that privatizing water utilities leads to unambiguous improvements in performance." A different study found no significant differences in efficiency between 28 public and 22 private W&S companies in 29 Pacific and Asian countries. A third study demonstrated that 2 private W&S companies were more efficient in comparison to 19 public W&S companies in sub-Saharan Africa. Small sample sizes, lack of data for important variables, data limited to a single year were all proposed as limitations to "isolating the relationship between privatization and efficiency." The findings are furthermore inconclusive as to whether W&S privatization contributes to environmental improvements or deteriorations. A different concern pertaining to privatization of the W&S sector is that such privatization could contribute to an increase in the monthly tariff for the provision of services. A substantial body of literature suggests that households in developing countries do not have access to improved W&S services due to the difficulty in obtaining an initial service connection rather than the inability to pay the service fee. "Different groups of customers often face substantially different service prices as a result of privatization." An additional concern regarding W&S

privatization is the possibility of "dramatic reduction in staff" associated with the transition from public to private W&S service provision.

Evidence reveals that private-sector participation (PSP) in relationship to the water and sanitation sector "will not benefit the majority of the...people who lack access to improved water supply and live in the world's poorest countries." General consensus exists regarding the pressing need to provide water access to 1.2 billion people in the world, water sanitation services to 2.4 billion people in the world, and to ameliorate current management of wastewater that has adverse effects on global freshwater and marine environments (Davis, 2005).

6.9 Poverty Reduction

An essential factor in reducing poverty is the access to affordable and safe drinking water (Morris & Cabera, 2003). The Humanitarian Charter defines water as "a basic human right". Most people who live in poverty identify the lack of access to safe water as one of the primary causes of poverty. The major beneficial impacts of safe water on poverty include improved health, productivity, school attendance, reduced spending on health care and reduced time fetching water. 53% of the rural and 17% of the urban sub-Saharan population lacked access to safe water in 2000. 35% to 50% of water systems in rural sub-Saharan Africa are not functioning (Harvey, 2008).

6.9.1 Poverty Reduction Strategy (PRS)

The World Bank and International Monetary Fund (IMF) established the Poverty Reduction Strategy (PRS) in 1999. As part of PRS, developing nations are mandated to complete a PRS paper (PRSP) as a means to access World Bank and IMF loans and Highly Indebted Poor Countries (HIPC) debt relief.

The document includes a description of the applicant country's macro-economic status, initiatives to promote growth and decrease poverty rates, societal and structural policies, primary sources of financing and related external financial necessities. 41 countries took part in PRS. 32 of member nations were in sub-Saharan Africa.

Generally, findings revealed that sub-Sahara African nations focused more attention on education and health care as opposed to the water and sanitation sector. "Access to water in rural areas is given low priority in reducing poverty." Education and health care sectors are thereafter expected to receive greater financial support compared to the water and sanitation sectors. 20 African countries have completed the PRSP process and 9 African countries have prepared interim PRSPs. 6 out of the 20 PRSPs or 30% had an adequate focus on water. The remaining 70% "did not have even a single paragraph specifically on water." 13 out of the 20 PRSPs or 65% had adequate focus on rural development, implying the greater importance on urban development. All 20 PRSPs placed sufficient focus on three major themes, including privatization, trade liberalization and decentralization. Trade liberalization relates to technology, supply chain and maintenance. Privatization has possible effects on financial and institutional challenges, technology, supply chain and maintenance. Finally, decentralization has possible effects on community and social challenges, monitoring, financial and institutional challenges. Of the 6 PRSPs with sufficient focus on water and sanitation sector, all highlighted the importance of community management in rural water service. Community management was the focus of 17 out of 20 PRSPs and therefore had emerged as the fourth major theme. Community management implies "empowering communities to take ownership of, and responsibility for, their own water supplies" (Harvey, 2008).

6.10 Environmental Injustice

Environmental injustice or "disproportionate environmental burdens by race and class" has been observed in the context of drinking water in the U.S. Findings from 5 years (2005-2010) of field data collected from California's San Joaquin Valley in the U.S. reveal elevated contamination levels of water resources at San Joaquin Valley. A direct link between race/ethnicity/socioeconomic status and contaminated drinking water supplied by non-compliant with federal standards community water systems was not established. However, it was found that "race and class are imbricated in almost all the factors," including natural, built and sociopolitical, "that have historically combined, and still combine, to produce this composite burden" (Balazs & Ray, 2014).

6.11 Water Governance

Water governance is defined as: "the range of political, social, economic and administrative systems that are in place to develop and manage water resources, and the delivery of water services, at different levels of society" (Franks & Cleaver, 2007).

The primary goal of the Clean Water Act (CWA) is to protect the waters in the U.S. Section 301 (a) of the CWA regulates the discharge of contaminants into water in the U.S. In 1972, the U.S. Congress passed amendments to the CWA with the Federal Water Pollution Control Act. The amendments intended "to restore and maintain the chemical, physical, and biological integrity of the Nation's waters". A permit system for regulating the sources of contamination was adopted (Fish, 2009). The U.S. CWA of 1972 prohibits the discharge of contaminants into U.S. navigable water without a National Pollution Discharge Elimination System (NPDES) permit. The CWA authorizes the EPA to issue such permits (Cabrera, 2014). However, due to a lack of contemporary legislation or rules, circuits continue to interpret the CWA on a case-by-case basis. The result is "extraneous consumption of court resources, inconsistent results, and decisions out-of-step with Congress' original intent in passing the Federal Water Pollution Control Act Amendments of 1972" (Fish, 2009).

The EPA has not been reliable in implementing the mandate. The EPA would often choose to not regulate pollutants. The agency has opted to grant NPDES exemptions "to avoid dealing with hard-to-regulate polluters (politically) and pollutants (physically)." "The pattern of EPA's actions, or lack thereof, has had little to do with clean water, and more to do with administrative or political convenience." It is recommended that Congress amends the law as a means to clarifying whether and to what degree should EPA regulate pollution sources which are currently in the "gray areas of the CWA" (Trott, 2009).

6.12 Public Participation

"Public and stakeholder participation is increasingly defended as a way to improve decisions, help build social capital, strengthen civil society and enhance the capability of communities to solve problems and pursue common concerns" (Videira et al., 2006).

6.13 Solution

There are a number of definitions of what sustainability is. According to the World Commission on Environment and Development (1987), "development to be sustainable must meet the needs of the present without compromising the ability of future generations to meet their own needs." (O'Keeffe, 2009). Intergenerational justice, founded by John Rawls, proposes that current generations have an ethical obligation to make certain that future generations are guaranteed fundamental "welfare" rights such as education, food, shelter and clean water (Goes et al., 2011). Sustainable management of water resources is a priority. "There are serious requirements for people to change their thinking and their habits at all levels" in order to establish sustainable management of water resources. It is increasingly being recognized that a multi-disciplinary approach is necessary in order to ensure sustainable water management (O'Keeffe, 2009).

"Water elemental endeavors to remember the shifting from one form into another."

By Sosé Gjelaj

Discussion

"Template for Extinction" provides additional clear evidence in support of Sosé's theory. Examples of problems created by governments and industries as a means to generate profit continue to expand. The deceit associated with such problems is highlighted in "Template for Extinction."

The "Leather Privilege" chapter revealed that children are employees of the industry. The "Conditions in Motion" and "A Sample of Inorganic Elements" chapters indicated that children's health is affected by industrial and chemical pollutants to a great extent. "Conditions in Motion" explained that contaminated non-operational arsenic dipping sites in South Africa were left unfenced which exposed children to the effects of intoxication. "Conditions in Motion" further elaborated upon the fact that screening of children at risk for Pb poisoning is only partially undertaken. Within the context herein, exposure to contaminated water contributes to nine out of ten deaths in children while the 1982 survey conducted by Levy et al. clearly demonstrated that dentists in the U.S. were generally prescribing fluoride supplementation to children without prior testing of fluoride levels children were

82

already exposed to from drinking water. Dentists were therefore prescribing fluoride tablets to children who may had not otherwise required additional fluoride intake. It is thus not a coincidence that 22% of surveyed children in the U.S. had developed fluorosis between 1986 and 1987 while 69% of children were estimated to have had developed fluorosis in the U.S. in 1998. A different study indicated that 78% of children in fluoridated communities suffered from fluorosis while 3% to 45% of children from non-fluoridated communities experienced fluorosis. Fluorosis leads to deformity of organs, gastrointestinal ailments, impaired growth and reduced uptake of calcium which automatically translates into profit for the health care industry.

The above examples solidify the conclusion that the only purpose of governments and industries is the creation of problems as a means to acquire financial gain at the expense of human health, including childhood well-being. Children are not being protected by industries and governments while opportunities for childhood disease are being created (i.e., problems such as fluoridation of water) in order for industries to profit from childhood impairments.

To add to the topic on fluoride, fluoridation is recommended as a preventative measure against dental caries, yet an obvious actual and not superficial solution to dental caries is the removal of factors leading to the development of dental caries. No such efforts are undertaken as healthy teeth and health in general signify zero profit for governments and corporations. Superficial solutions are devised to mask the root of the problem.

Further, the Supreme Court in Illinois agreed that the risk of fluoridation of drinking water is so drastic that fluoridation is unreasonable, yet allowed for the legal continuation of fluoridation of drinking water in the state of Illinois as "the question was shown to be debatable." Despite such risk assessment, fluoridation of drinking water is endorsed by supposedly trustworthy U.S. medical and scientific organizations which is a clear indication that fluoridation of water is a purposefully-designed problem, perpetuating the resulting governmental and industrial profit from human disease.

On a similar note, the EPA is responaible for regulating certain aspects of water pollution and the Ministry of Health in Peru is to be conducting proper monitoring of arsenic in drinking water when it is charged to do so. Yet, the above mentioned governmental institutions do not accomplish such tasks as the purpose is to pollute the environment so profit can be acquired, not prevent or solve pollution. What other reasonable explanation could there be for the lack of intervention? The WHO also removed the manganese standard in drinking water when manganese is still prevalent in drinking water in certain parts of the world. The intention is clearly not to prevent contamination, but rather to poison humanity in the name of profit from resulting disease. The WHO, EPA and the Ministry of Health are, similarly to U.S. medical and scientific organizations as part of the fluoride example, the mask beneath which the deceit takes place. The very existence of such organizations and their very names (i.e., deceit) reassure humanity that problems are being attended to by governmental agencies while, in actuality, they are being created by the government for purposes of financial gain.

Along the same line of reasoning, while chlorine dioxide poses health hazards, iodine is a safe water disinfectant that also prevents thyroid disorders. WHO and the EPA reject the use of iodine as a water disinfectant under the arguments that there is an ill-defined potential that iodine affects the thyroid and that there are alternative water treatment methods. In terms of the first argument, chlorine dioxide poses scientifically established health hazards, yet it is not being removed as a water disinfectant. Iodine's adverse health effects have not even been demonstrated and yet it is not being considered as a water disinfectant. Research articles hint to the high prevalence of thyroid disorder and the effectiveness of iodine as a treatment modality. The loss of profit for health care industries if thyroid problems were prevented on a global level with the use of iodine as a water disinfectant can only be imagined. The profit decreases even further when the adverse health effects resulting from chlorine disinfectants are removed from the equation. In terms of the second argument, the availability of alternative water treatment methods does not guarantee effectiveness and does not automatically disqualify the effectiveness of other water disinfectants.

Moreover, the article on surface water contamination with road salts in the GTA provides support for Sosé's theory that truth is hidden from humanity (i.e., "gathering salt usage data from the municipality officials proved very challenging"). There would be no reason to hide the truth unless revealing it would not only expose governmental and industrial actual motives for generating profit through the creation of problems, but will also contribute to public awareness of the environmental damage resulting from the use of road salts. Such awareness could lead to limited use of road salts and/or the implementation of safer alternative methods thus reducing opportunities for profit.

Last, but not least, water privatization has become a global epidemic. Not only a few companies currently own the fresh water sector worldwide, but while private water companies continue to profit from rising water tariffs, water access and service is steadily deteriorating. The theory that problems are created in order to acquire not only profit, but also power, is beginning to take shape with the water privatization example. The order appears to be as follows: no capital, no access to water, human exploitation and/or demise, absolute power.

References

Amirsalari, F., Li, J., Guan, X. & Booty, W. G. (2013). Investigation of correlation between remotely sensed impervious surfaces and chloride concentrations. *International Journal of Remote Sensing, 34*(5), 1507-1525.

Bagh, B., Das, K. & Ray, S. (2006). Endemic fluorosis through drinking water and its remedial measure by bacterial population – a mathematical model. *Journal of Biological Systems, 14*(1), 31-41.

Backer, H. & Hollowell, J. (2000). Use of iodine for water disinfection: Iodine toxicity and maximum recommended dose. *Environmental Health Perspectives, 108*(8), 679-684.

Balazs, C. L. & Ray, I. (2014). The drinking water disparities framework: On the origins and persistence of inequities in exposure. *American Journal of Public Health, 104*(4), 603-611.

Benbrahim-Tallaa, L. & Waalkes, M. P. (2008). Inorganic arsenic and human prostate cancer. *Environmental Health Perspectives, 116*(2), 158-164.

Cabrera, N. (2014). Plan meaning or pragmatics? Differing interpretations of the Clean Water Act's jurisdictional provisions. *Boston College Environmental Affairs Law Review, 41*, 1-13.

Chakraborty, T. & De, M. (2009). Clastogenic effects of inorganic arsenic salts on human chromosomes in vitro. *Drug and Chemical Toxicology, 32*(2), 169-173.

Chen, G.-H. & Wong, M.-T. (2004). Impact of increased chloride concentration on nitrifying-activated sludge cultures. *Journal of Environmental Engineering, 130*(2), 116-125.

Cheng, Q., Wu, H., Wu, Y., Li, H., Zhang, X. & Wang, W. (2013). Groundwater quality and the potentiality in health risk assessment in Zhengzhou, China. *Aquatic Ecosystem Health & Management, 16*(1), 94-103.

Chenoweth, J. & Bird, J. (2003). GMI theme issue: The business of water and sustainable development. *Greener Management International, 42*, 5-8.

Christoffel, T. (1985). Fluorides, facts and fanatics: Public health advocacy shouldn't stop at the courthouse door. *American Journal of Public Health, 75*(8), 888-891.

Clark, R. M. & Sivaganesan, M. (2002). Predicting chlorine residuals in drinking water: Second order model. *Journal of Water Resources Planning and Management, 128*(2), 152-161.

Dagdeviren, H. (2008). Waiting for miracles: The commercialization of urban water services in Zambia. *Development and Change, 39*(1), 101-121.

Davey, J. C., Nomikos, A. P., Wungjiranirun, M., Sherman, J. R., Ingram, L., Batki, C., Lariviere, J. P. & Hamilton, J. W. (2008). Arsenic as an endocrine disruptor: Arsenic disrupts retinoic acid receptor-and thyroid hormone receptor-mediated gene regulation and thyroid hormone-mediated amphibian tail metamorphosis. *Environmental Health Perspectives, 116*(2), 165-172.

Davis, J. (2005). Private-sector participation in the water and sanitation sector. *Annual Review of Environment & Resources, 30*, 145-183.

der Mude, A. V. (2011). Endocrine-disrupting chemicals: Testing to protect future generations. *Environmental Affairs, 38*, 509-535.

Dieperink, C. (2011). International water negotiations under asymmetry, lessons from the Rhine chlorides dispute settlement (1931-2004). *International Environmental Agreements: Politics, Law & Economics, 11*, 139-157.

Dong, Y., Kim, J. & Lewis, G. D. (2010). Evaluation of methodology for detection of human adenoviruses in wastewater, drinking water, stream water and recreational waters. *Journal of Applied Microbiology, 108*, 800-809.

Erdal, S. & Buchanan, S. N. (2005). A quantitative look at fluorosis, fluoride exposure, and intake in children using a health risk assessment approach. *Environmental Health Perspectives, 113*(1), 111-117.

Fisch, A., Pichard, E., Prazuck, T., Sebbag, R., Torres, G., Gemez, G. & Gentilin, M. (1993). A new approach to combatting iodine deficiency in developing countries: The controlled release of iodine in water by a silicone elastomer. *American Journal of Public Health, 83*(4), 540-545.

Foo, L.-C., Mahmud, N. & Satgunasingam, N. (1998). Eliminating iodine deficiency in rural Sarawak, Malaysia: The relevance of water iodization. *American Journal of Public Health, 88*(4), 680-681.

Franks, T. & Cleaver, F. (2007). Water governance and poverty: A framework for analysis. *Progress in Development Studies, 7*(4), 291-306.

Frisbie, S. H., Mitchell, E. J., Dustin, H., Maynard, D. M. & Sarkar, B. (2012). World Health Organization discontinues its drinking-water guideline for manganese. Environmental Health Perspectives, *120*(6), 775-778.

Garcia-Prieto, J. C., Cachaza, J. M., Perez-Galende, P. & Roig, M. G. (2012). Impact of drought on the ecological and chemical status of surface water and on the content of arsenic and fluoride pollutants of groundwater in the province of Salamanca (Western Spain). *Chemistry and Ecology, 28*(6), 545-560.

Gatseva, P. D. & Argirova, M. D. (2005). Iodine status of children living in areas with high nitrate levels in water. *Archives of Environmental & Occupational Health, 60*(6), 317-319.

George, C. M., Sima, L., Arias, M. H. J., Mihalic, J., Cabrera, L. Z., Danz, D., Checkley, W. & Gilman, R. H. (2014). Arsenic exposure in drinking water: An unrecognized health threat in Peru. *Bulletin of the World Health Organization, 92*, 565-572.

Gupta, R. S., Khan, T. I., Agrawal, D. & Kachhawa, J. B. S. (2007). The toxic effects of sodium fluoride on the reproductive system of male rats. *Toxicology and Industrial Health, 23*, 507-513.

Hafeman, D., Factor-Litvak, P., Cheng, Z., van Geen, A. & Ahsan, H. (2007). Association between manganese exposure through drinking water and infant mortality in Bangladesh. *Environmental Health Perspectives, 115*(7), 1107-1112.

Hamner, S., Tripathi, A., Mishra, R. K., Bouskill, N., Broadway, S. C., Pyle, B. H. & Ford, T. E. (2006). The role of water use patterns and sewage pollution in incidence of water-borne/enteric diseases along the Ganges River in Varanasi, India. *International Journal of Environmenta Health Research, 16*(2), 113-132.

Hernandez-Zavala, A., Valenzuela, O. L., Matousek, T., Drobna, Z., Dedina, J., Garcia-Vargas, G. G., Thomas, D. J., Del Razo, L. M & Styblo, M. (2008). Speciation of arsenic in exfoliated urinary bladder epithelial cells from individuals exposed to arsenic in drinking water. *Environmental Health Perspectives, 116*(12), 1656-1660.

Hoddinott, J., Cohen, M. J. & Barrett, C. B. (2008). Renegotiating the food aid convention: Background, context, and issues. *Global Governance, 14,* 283-304.

Hosni, A. A., Shane, W. T., Szabo, J. G. & Bishop, P. L. (2009). The disinfection efficacy of chlorine and chlorine dioxide as disinfectants of *Bacillus globigii*, a surrogate for *Bacillus anthracis*, in water networks: A comparative study. *Canadian Journal of Civil Engineering, 36,* 732-737.

Hrudey, S. E., Bull, R. J., Cotruvo, J. A., Paoli, G. & Wilson, M. (2013). Drinking water as a proportion of total human exposure to volatile *N*-nitrosamines. *Risk Analysis, 33*(12), 2179-2208.

Huby, M. & Stevenson, S. (2003). Meeting need and achieving sustainability in water project interventions. *Progress in Development Studies, 3*(3), 196-209.

Hughes, M. F. (2006). Biomarkers of exposure: A case study with inorganic arsenic. *Environmental Health Perspectives, 114* (11), 1790-1796.

Inyang, H. I. (2004). Geochemical and health policy dimensions of the controversy about fluorides in water resources. *Journal of Environmental Engineering, 130*(2),

113-114.

Kamaluddin, M. & Zwiazek, J. J. (2003). Fluoride inhibits root water transport and affects leaf expansion and gas exchange in aspen (*Populus tremuloides*) seedlings. *PhysiologiaPlantarum, 117*, 368-375.

Kordella, S., Geraga, M., Paptheodorou, G., Fakiris, E. & Mitropoulou, I. M. (2013). Litter composition and source contribution for 80 beaches in Greece, Eastern Mediterranean: A nationwide voluntary clean-up campaign. *Aquatic Ecosystem Health & Management, 16*(1), 111-118.

Kreisberg, J. (2007). Greener pharmacy: Proper medicine disposal protects the environment. *Integrative Medicine, 6*(4), 50-52.

Kurttio, P., Komulainen, H., Leino, A., Salonen, L., Auvinen, A. & Saha, H. (2005). Bone as a possible target of chemical toxicity of natural uranium in drinking water. *Environmental Health Perspectives, 113*(1), 68-72.

Leurs, L. J., Schouten, L. J., Mons, M. N., Goldbohm, R. A., van den Brandt, P. A. (2010). Relationship between tap water hardness, magnesium, and calcium concentration and mortality due to ischemic heart disease or stroke in the Netherlands. *Environmental Health Perspectives, 118*(3), 414-420.

Levy, S. M., Bawden, J. W., Bowden, B. S. & Rozier, R. G. (1984). Fluoride analyses of patient water supplies requested by North Carolina health professional. *American Journal of Public Health, 74*(12), 1412-1414.

Li, H., Campbell, A., Ali, S. F., Cong, P. & Bondy, S. C. (2007). Chronic exposure to low levels of aluminum alters cerebral cell signaling in response to acute MPTP administration. *Toxicology and Industrial Health, 22*, 515-524.

Mahajan, R. K., Walia, T. P. S., Lark, B. S. & Sumanjit, (2006). Analysis of physical and chemical parameters of bottled drinking water. *International Journal of Environmental Health Research, 16*(2), 89-98.

Macek, M. D., Matte, T. D., Sinks, T. & Malvitz, D. M. (2006). Blood lead concentrations in children and method of water fluoridation in the United States, 1988-1994. *Environmental Health Perspectives, 114*(1), 130-134.

Makhado, R., Bologo, T. & Knight, R. (2006). Fluoride removal from rural spring water using wood ash. *South African Journal of Science, 102* (5/6), 272.

Martinelli, S. S., Cavalli, S. B., Pires, P. P., Proenca, L. C. & Da Costa Proenca, R. P. (2012). Water consumption in meat thawing under running water: Sustainability in meal production. *Journal of Culinary Science & Technology, 10,* 311-325.

Michael, G. E., Miday, R. K., Bercz, J. P., Miller, R. G., Greathouse, D. G., Kraemer, D. F. & Lucas, J. B. (1981). Chlorine dioxide water disinfection: A prospective epidemiology study. *Archives of Environmental Health, 36*(1), 20-27.

Morris, L. & Cabera, L. F. G. (2003). The Involvement of the private sector in water servicing: Effects on the urban poor in the case of Aguascalientes, Mexico. *Greener Management International, 42,* 35-46.

Mulamattathil, S. G., Bezuidenhout, C. & Mbewe, M. (2014). Biofilm formation in surface and drinking water distribution systems in Mafikeng, South Africa. *South African Journal of Science, 110*(11/12), 81-89.

Nieuwenhuijsen, M. J., Toledano, M. B., Bennett, J., Best, N., Hambly, P., de Hooogh, C., Wellesley, D., Boyd, P. A., Abramsky, L. Dattani, N., Fawell, J., Briggs, D., Jarup, L. & Elliott, P. (2008). Chlorination disinfection by-products and risk of congenital anomalies in England and Wales. *Environmental Health Perspectives, 116*(2), 216-222.

Patel, A. I., Hecht, K., Hampton, K. E., Grumbach, J. M., Braff-Guajardo, E. & Brindis, C. D. (2014). Tapping into water: Key considerations for achieving excellence in school drinking water access. *American Journal of Public Health, 104*(7), 1314-1319.

Pi, J., Yamauchi, H., Sun, G., Yoshida, T., Aikawa, H., Fujimoto, W.,

Iso, H., Cui, R., Waalkes, M. P. & Kumagai, Y. (2005). Vascular dysfunction in patients with chronic arsenosis can be reversed by reduction of arsenic exposure. *Environmental Health Perspectives, 113*(3), 339-341.

Pillon, A., Boussioux, A.-M., Escande, A., Ait-Aissa, S., Gomez, E., Fenet, H., Ruff, M., Moras, D., Vignon, F., Duchesne, M.-J., Casellas, C., Nicolas, J.-C. & Balaguer, P. (2005). Binding of estrogenic compounds to recombinant estrogen receptor-α: Application to environmental analysis. *Environmental Health Perspectives, 113*(3), 278-284.

Popkin, B. M., D'Anci, K. E. & Rosenberg, I. H. (2010). Water, hydration, and health. *Nutrition Reviews, 68*(8), 439-458.

Qin, X., Yang, B., Gao, F. & Chen, G. (2013). Treatment of restaurant wastewater by pilot-scale electrocoagulation-electroflotation: Optimization of operating conditions. *Journal of Environmental Engineering, 139*(7), 1004-1016.

Rasanen, N. H. J., Rintala, H., Miettinen, I. T. & Torvinen, E. (2013). Comparison of culture and qPCR methods in detection of mycobacteria from drinking waters. *Canadian Journal of Microbiology, 59*, 280-286.

Rutjes, S. A., Lodder, W. J., van Leeuwen, A. D. & de Roda Husman, A. M. (2009). Detection of infectious rotavirus in naturally contaminated source waters for drinking water production. *Journal of Applied Microbiology, 107*, 97-105.

Sharma, A., Gupta, M., & Shanker, A. (2008). Fenvalerate residue level and dissipation in tea and in its infusion. *Food Additives and Contaminants, 25*(1), 97-104.

Shou, L., Zeng, J., Liao, Y., Xu, T., Gao, A., Chen, Z., Chen, Q. & Yang, J. (2013). Temporal and spatial variability of benthic macrofauna communities in the Yangtze River estuary and adjacent area. *Aquatic Ecosystem Health & Management, 16*(1), 31-39.

Schuhmacher-Wolz, U., Dieter, H. H., Klein, D. & Schneider, K. (2009). Oral exposure toinorganic arsenic: Evaluation of its carcinogenic and non-carcinogenic effects. *Critical

Reviews in Toxicology, 39(4), 271-298.

Trott, A. (2009). Lack of deference: The North Cicuit's misstep in *NRDC v EPA. Ecological Law Quarterly, 36,* 355-379.

Tudor, D. T. & Williams, A. T. (2008). Important aspects of beach pollution to managers: Wales and the Bristol Channel, UK. *Journal of Coastal Research, 24*(3), 735-745.

Uejio, C. K., Yale, S. H., Malecki, K., Borchardt, M. A., Anderson, H. A. & Patz, J. A. (2014). Drinking water systems, hydrology, and childhood gastrointestinal illness in Central and Northern Wisconsin. *American Journal of Public Health, 104*(4), 639-646.

Usio, N., Imada, M., Nakagawa, M., Akasaka, M. & Takamura, N. (2013). Effects of pond draining on biodiversity and water quality of farm ponds. *Conservation Biology, 27*(6), 1429-1438.

Wang, Y. & Zhou, J. (2013). Endocrine disrupting chemicals in aquatic environments: A potential reason for organism extinction? *Aquatic Ecosystem Health & Management, 16*(1), 88-93.

Wasserman, G. A., Liu, X., Parvez, F., Ahsan, H., Levy, D., Factor-Lityak, P., Kline, J., van Geen, A., Slavkovich, V., LoIacono, N. J., Cheng, Z., Zheng, Y. & Graziano, J. H. (2006). Water manganese exposure and children's intellectual function in Araihazar, Bangladesh. *Environmental Health Perspectives, 114*(1), 124-129.

Wimberger, S. (2009). Consideration of alternatives in environmental impact reports: The   importance of CEQAs procedural requirements. *Ecological Law Quarterly, 36,* 499-524.

Yakub, I. & Soboyejo, W. (2013). Adsorption of fluoride from water using sintered clay-hydroxyapatite composites. *Journal of Environmental Engineering, 139*(7), 995-1003.

Yuling, S. & Lein, H. (2010). Treating water as an economic good: Policies and practices in irrigation agriculture in Xinjiang, China. *The Geographical Journal, 176*(2), 124-137.

Zoller, U. (2006). Water reuse/recycling and reclamation in semiarid zones: The Israeli case of salination and "hard" surfactants pollution of aquifers. *Journal of Environmental Engineering, 132*(6), 683-688.

Chapter 7

GENERA

Overview

"Genera" generally elaborates on the effects of pollution on aquatic species. Policies pertaining to the protection of marine mammal species are presented.

7.1 Coral Reefs, China

Coral reefs have a significant conservation value. Coral reefs have declined by 80% in the past 30 years alongside the Chinese mainland and adjacent Hainan Island. Pollution, coastal development, overfishing and damaging fishing practices followed by climate change have contributed to the accelerated decline of coral reefs in China. In other words, the rapid decline of coral reefs is caused by "a failure of governance." Governance is defined as "the structures and processes by which multiple actors...share power and make decisions." Governance consists of formal components such as laws and regulations and informal components such as religion, media, public opinion and commerce. "These informal elements are powerful. Consequently, societal norms and attitudes in China are major aspects of the wicked problem of sustaining coral reefs. From this broader perspective, governance in China is not only top-down but also bottom-up." Approximately 57 marine protected areas (MPAs) have been established in China since 1990. "However, these parks are small and do not adequately represent the diversity of China's coastal ecosystems." Furthermore, there is no mandatory environmental monitoring system of these areas, resulting in "unrecognized and unreported...change such as the rapid loss of coral cover" (Hughes et al., 2012).

7.2 Amphibian Species

7.2.1 Whale

Marine spatial planning (MSP) has established a framework for the management of the marine environment (i.e., military training; fishing). "MSP must be based on ecological principles to sustain ecosystem integrity." Large whales are susceptible to collision with vessels of various size, type and class. The United States (U.S.) Marine Mammal Protection Act regulates the number of aquatic species that "may be removed" on a yearly basis by "anthropogenic causes" while ensuring that the species sustain optimal population levels. Optimum sustainable population is defined as "the number of animals which will result in the maximum productivity of the population or the species, keeping in mind the carrying capacity of the habitat and the health of the ecosystem of which they form a constituent element." Three blue whales along the U.S. West Coast are permitted to be removed by anthropogenic causes annually.

The loss of blue whales by ship strikes along the coast of California has been documented for approximately twenty years. Concerns were expressed in 2007 as "at least" 4 blue whale individuals were struck by ships off Southern California. An average of 1.8 blue whales, 1.2 fin whales and 1 humpback whale were reported as being killed annually by ship strikes along the coast of California between 2005 and 2010. Such documented number is considered an "underestimate of the actual number of strikes" due to the fact that "ship strikes have low probability of detection." A more realistic estimate indicates that 10.6 blue whales, 7.1 fin whales and 5.9 humpback whales are killed by ships every year along the California coast. Ocean contamination, climate change, noise from commercial shipping activities and entanglement in fishing equipment are additional probable "threats to large whales" along the California Current (Redfern et al., 2013).

7.3 Pollution

7.3.1 Endocrine Disrupting Contaminants (EDCs)

Declines of amphibian species have been demonstrated in "many parts of the world, which is a concern" as amphibian species play a significant role as grazers, prey species, and predators in aquatic as well as terrestrial ecosystems. One possible reason for the decline of amphibian species is exposure to anthropogenic endocrine disrupting contaminants such as PCBs, p,p-dichlorodiphenyltrichloroethane (DDT) and polycyclic aromatic hydrocarbons (PAHs). PAHs, "widely disseminated endocrine disruptors," are emitted during the combustion of coal, oil, gas, wood and waste. DDT is an organochlorine pesticide that has been found to demasculanize birds. The widespread herbicide atrazine has been found to compromise amphibian reproduction and/or development.

The proportion of intersex gonads in Illinois cricket frog (Acris crepitans) samples deposited at a museum in Illinois was investigated. Findings from the study revealed that intersex gonads in cricket frogs increased parallel to industrial growth and the introduction of PCBs between 1930 and 1946 and peaked during the highest production of PCBs and DDT between 1946 and 1959. Increased public and environmental regulations contributed to the reduction of PCB and DDT use in the Midwestern U.S. between 1960 and 1979. Intersex gonads in cricket frogs began to decline since 1960 and are currently at a "near-baseline" state. A strong association between intersex and endocrine disruptors was therefore demonstrated (Reeder et al., 2005).

7.3.2 Mercury (Hg)

Building 615 at Fort Totten in New York State is a former U.S.

army site. Building 615 released Hg which "most likely" has been "used in switches for torpedoes and mines" via a drainpipe into Little Bay. Once released into Little Bay, Hg "was deposited in the sediment." Hg can be present in either dissolved or particulate form in aquatic systems. Hg emitted in the air can be absorbed by terrestrial or aquatic plants and inhaled by animals. Aquatic plants and animals can be directly exposed to Hg present in water. Hg in drinking water can be consumed by terrestrial animals. There is a proposal to convert 50 acres of Fort Totten into a public recreational area. An ecological risk assessment (ERA) of the region demonstrates that the "mercury spill" would pose "little to no risk to the ecological receptors evaluated" (i.e., Little Bay aquatic species; regional mammal and bird species) (Goldblum et al., 2013).

7.3.3 Bacteria

The exoskeletons of arthropods and aquatic crustaceans consist of the biopolymer, chitin. Bacterial damage to the exoskeleton contributes to chitin impairment and causes shell disease syndrome in a variety of crustaceans. Shell disease is considered a syndrome as there are more than one causing factors. Shell disease is characterized by the development of "black-spot lesions" on the exoskeleton of crustaceans. The lesions degrade or break the chitin, penetrating into the soft tissue underneath the exoskeleton and causing mortality in worst cases. "Anthropogenic disturbances play a critical role in disease progression." Substantial evidence reveals a strong association between shell disease syndrome, metal ions or insecticides pollution while sewage pollution is a probable though not yet proven contributing factor (Vogan et al., 2008).
　　　　Pathogen indicator bacteria is the primary cause of the deterioration of water quality in U.S. rivers and streams. There are a number of gaps in the knowledge base regarding environmental faecal microbe dynamics which prevents the

development of microbial pollution models. "Uncertainty in the modeling of microbial pollution from agriculture is, perhaps, an interesting challenge for scientists but an inconvenient truth for policy-makers" (Oliver et al., 2009).

Discussion

Deriving from Sosé's theory, the U.S. Marine Mammal Protection Act is yet another mask beneath which problems are being orchestrated. The name itself implies that action is taken to protect marine mammals. However, the reality is exactly the opposite. Action is being taken to allow for the demise of mammals, not to protect them. The deceit is clear. What is not clear from reviewed research on the topic is how exactly is the creation of such a problem profiting governments and industries. The possibilities are unlimited. Similarly, the opportunities for profit presenting themselves as a result of the pollution of aquatic environments are countless (i.e., treatment of disease; fundraising).

References

Goldblum, D. K., Cora, M. G. & Rak, A. (2013). Assessing ecological risk from mercury in Little Bay estuary: A study from Fort Totten in New York. *Environmental Quality Management, 23*(1), 23-41.

Hughes, T. P., Huang, H. & Young, M. A. L. (2012). The wicked problem of China's disappearing coral reefs. *Conservation Biology, 27*(2), 261-269.

Oliver, D. M., Heathwaite, A. L., Fish, R. D., Chadwisk, D. R., Hodgson, C. J., Winter, M. & Butler, A. J. (2009). Scale appropriate modelling of diffuse microbial pollution from agriculture. *Progress in Physical Geography, 33*(3), 358-377.

Reeder, A. L., Ruiz, M. O., Pessier, A., Brown, L. E., Levengood, J. M., Philipps, C. A., Wheeler, M. B., Warner, R. E. & Beasley,

V. R. (2005). Intersexuality and the cricket frog decline: Historic and geographic trends. *Environmental Health Perspectives, 113*(3), 261-265.

Redfern, J. V., McKenna, M. F., Moore, T. J., Calambokidis, J., Deangelis, M. L., Becker, E. A., Barlow, J., Forney, K. A., Fiedler, P. C. & Chivers, S. J. (2013). Assessing the risk of ships striking large whales in marine spatial planning. *Conservation Biology, 27*(2), 292-302.

Vogan, C. L., Powell, A. & Rowley, A. F. (2008). Shell disease in crustaceans – just chitin recycling gone wrong? *Environmental Microbiology, 10*(4), 826-835.

Chapter 8

FISH EXHAUSTION

Overview

"Fish Exhaustion" describes the current state of wild fish populations and elaborates on the prevalence and environmental hazards pertaining to aquaculture. Fish slaughter techniques are described and fish welfare issues are discussed. The role of governments and industry in aquaculture is presented.

8.1 Health Risks

8.1.1 Prostate Cancer

Tortfadottir et al., (2013) investigated the relationship between fish consumption and prostate cancer in men. Very high consumption of fish during the early and midlife years was not found to be associated with prostate cancer risk. Smoked or salted fish was found to increase prostate cancer risk in early and later life.

8.2 Contamination

The average daily consumption of fish in the United States (U.S.) general population is 15.65 g/day of consumed fish or 20.1 g/day of uncooked fish (Kissinger et al., 2010). "Despite the need for such assessments, there is currently no accepted method to compare net risks with net benefits of consuming fish" (Turyk et al., 2012). There is a safety concern regarding fish and seafood consumption due to the presence of contaminants such as heavy metals in fish seafood products (Pastorelli et al., 2012). Concerns have been expressed regarding the safety of non-commercial fish and the positive relationship between fish contamination and neurobehavioral impairments in children (Burger, 2002). Higher fish intake leads to greater risk of toxicity due to fish contamination by pollutants such as methyl mercury (Oken et al., 2012).

8.2.1 Methyl Mercury (MeHg) & Polychlorinated Biphenyls (PCBs)

Human beings are exposed to persistent organic pollutants (POPs) primarily through the consumption of contaminated food such as fish. PCBs represent an example of POPs. PCBs are contained in fish (Binnington et al., 2014).

PCBs and MeHg contaminant levels are high in some types of fish, contributing to adverse health effects in human beings. The only substantial intake of MeHg is derived from fish consumption. A positive correlation has been established between MeHg and PCB concentrations in fish, maternal fish consumption and neurobehavioral impairments in children. Women consuming large quantities of contaminated fish from Lake Ontario have been found to experience declines in fecundity (Burger, 2002).

Discontinuing or reducing fish consumption during pregnancy is "largely ineffective" in "reducing prenatal, postnatal, and childhood exposures" to pollutants due to extended elimination half-lives of certain POPs such as PCBs (Binnington et al., 2014).

Great Lakes waters and fish have been contaminated with chemicals originating from urbanization, agriculture, transportation and industry. Such Great Lakes pollutants include PCBs, MeHg and chlorinated pesticides among others (Turyk et al., 2012).

Contaminated fish and marine mammals pose greater health risks to Native American and Canadian indigenous people as they tend to consume greater qualities of seafood compared with the general population (Kissinger et al., 2010).

8.2.2 Advisories

An accelerating number of lakes and rivers in the U.S. are under consumption advisories particularly in relation to Hg, PCBs, dioxins, chlordane and other heavy metals. 16% of the lakes in the U.S. are under consumption advisories (Burger, 2002). The European Union (EU) established maximum levels (MLs) for Hg, Cd and Pb in fish and seafood produce (Pastorelli et al., 2012).

"People are faced with conflicting information about the risks and benefits of consuming fish." State agencies in the U.S. issue advisories in relation to the risks of consuming contaminated fish. Such advisories particularly in regards to Hg and PCBs in fish have increased in the past decade. However, very few advisories concern commercially purchased fish (Burger, 2002).

8.2.3 Labeling

Accurate labeling is essential in allowing the public to make informed choices of their fish consumption patterns. However, many types of fish from various parts of the world are "sold under a common rubric such as tuna or flounder." A study analyzing the molecular content of fish sold as red snapper revealed that 45% of the sold fish was actually red snapper fish (Burger, 2002). Furthermore, labels on fish products are not designed to "deal with the plurality of moral views on why sustainability/welfare is morally relevant" (Kalshoven & Meijboom, 2013).

8.3 Wild Fish

"As the world population is growing and government directives tell us to consume more fatty acids by consuming more fish, the demand for fish is increasing" (Bovenkerk &

Meijboom, 2013). Seafood provides about 115 million tons of "human food" annually. The per capita annual consumption of seafood on a global level is c. 17 kg (live weight equivalent). 80 million tons of the world's supply of seafood per year is derived from wild capture sea fisheries (Mitchell, 2011). "Stocks of wild fish are not adequate to meet the nutrient demands of the growing world population" (Oken et al., 2012). "A number of the world's fisheries are currently depleted through overfishing" (Mitchell, 2011). It has been estimated that "commercial fishing has wiped out 90% of large fish such as swordfish, cod, marlin, and sharks." Marine environments and fisheries are exposed to additional "serious threats" such as invasion of exotic species, pollution, climate change, habitat alteration and coastal developments. Commercial fishing, the primary cause of declining fisheries, has compromised the resilience or the ability of fisheries and marine ecosystems to "withstand other amounting environmental pressures" (Naylor & Burke, 2005).

Traditional Command and Control (CAC) fishery regulation instruments such as limit on catch and number of boats, closed seasons and regulations regarding fishing technology have not been effective in preventing wild fish over-exploitation. This is due to "high and often prohibitive costs of control in open-access conditions" and taxes which are "generally unpopular and not politically feasible especially in developing countries" (Madhoo, 2011).

8.4 Law

"In law, the animal issue is often addressed in terms of legal subject/object categories. The evolution of modern Western societies in their relationship to animals as well as scientific progress, especially in genetics and ethology, show animals to be very similar to man, hardly even a distant cousin, while law insists on considering them as foreign entities, in other words, as things" (Coutellec & Doussan, 2012).

8.5 Aquaculture

The human population is growing exponentially and so is the demand for fish (Bergqvist & Gunnarsson, 2013). The consumption of fish has increased two-fold since the 1970s (Naylor & Burke, 2005). Over-fishing or unsustainable fishing practices are contributing to a decline of wild fish populations. The decline of wild fish populations poses environmental challenges as fish in the world's oceans "play a vital protective role of their ecosystems" (Bergqvist & Gunnarsson, 2013). The over-exploitation of wild fish populations (Bovenkerk & Meijboom, 2013) in addition to the increase in demand for fish (Bergqvist & Gunnarsson, 2013) have contributed to the development of aquaculture (Bovenkerk & Meijboom, 2013) in past decades (Bergqvist & Gunnarsson, 2013). "Aquaculture is the fastest growing animal-production sector in the world" (Bovenkerk & Meijboom, 2013).

9% of the total fish consumed in the 1980s was derived from farms (Bergqvist & Gunnarsson, 2013). Aquaculture supplied 43% of fish for human consumption in 2007 "and these numbers appear to be increasing" (Bovenkerk & Meijboom, 2013). Many of the depleted wild fish species such as Atlantic cod, Atlantic halibut, mutton snapper and bluefish tuna are farmed in cages or net pens at the bottom of oceans (Naylor & Burke, 2005).

Aquaculture industry is not meeting the demand for fish and neither is it meeting environmental needs. "The obvious environmental and animal welfare aspects of finfish aquaculture make it hard to ethically defend a fish diet" (Bergqvist & Gunnarsson, 2013).

8.5.1 Welfare

Human beings have historically created strong emotional attachments with animals other than fish. Emotional bonds with fish are difficult to make due to the air-water boundaries, challenges in communicating with fish, our inability to interpret the facial expressions and body language of fish, the lack of sound to indicate their psychological state. Despite the fact that fish do not withdraw feelings of empathy in human beings, "does this mean that it is right to exclude them from moral concern?"

Fish is being recognized as probable sentient animals and the public is increasingly being concerned about fish welfare in aquaculture (Bergqvist & Gunnarsson, 2013). There are two differing perspectives regarding fish welfare. On the one hand, people are aware that fish perceives pain which may prevent fish consumption, yet "an exception is often made when it concerns fish." On the other hand, people are not convinced that fish experiences pain which contributes to fish being treated in a different way compared to other animals (Bovenkerk & Meijboom, 2013). Recent studies demonstrate that finfish contain nociceptors which are receptors able to perceive pain. If fish can experience pain, then they are to be regarded as sentient creatures. Yet, there is a debate as to whether fish "have an actual cognitive experience of pain" (Bergqvist & Gunnarsson, 2013). Weighing, defining and measuring fish welfare is essential in "implementing welfare standards" (Bovenkerk & Meijboom, 2013). Fish welfare is potentially compromised during the various stages of production including breeding, growth, capturing, transportation and slaughter (Bergqvist & Gunnarsson, 2013).

Fish are captured primarily to be transported to pens for growth or to slaughter. Fish capturing and transportation can cause a number of potential "stressors for the animals." Poor water quality with low concentrations of oxygen, crowding, exhaustion due to the experience of multiple stressors, food deprivation often for several days or even weeks, physical harm such as skin abrasions during handling, increased aggression resulting from starvation as well as the spread of disease are examples of such potential stressors

(Bergqvist & Gunnarsson, 2013).

8.5.2 Slaughter Methods

Slaughter methods are "often adjusted to assure efficiency, not fish welfare." Indifferently of the technique being used, slaughter is highly stressful for farmed fish. Asphyxia, ice stunning, exsanguination, electrocution, carbon dioxide stunning, percussive stunning, spiking, salt bath combined with evisceration also referred to as gutting are the fish slaughter techniques currently used in Europe (Bergqvist & Gunnarsson, 2013).

8.5.2.1 Asphyxia

Asphyxia, the oldest fish slaughter technique, basically means suffocation resulting from taking fish out of the water. Asphyxia is characterized by "high occurrence of stress reaction in the animals". It takes between 5 minutes and longer than one hour for the fish to lose consciousness once it is taken outside of the water (Bergqvist & Gunnarsson, 2013).

8.5.2.2 Ice Stunning

Ice stunning essentially means the immersion of fish in ice water. The cause of death is anoxia. It takes between 12 minutes and 198 minutes for the fish to lose consciousness following immersion. This technique is described as "inhumane" (Bergqvist & Gunnarsson, 2013).

8.5.2.3 Exsanguination

Exsanguination which involves death through gill cutting without prior stunning has been banned as it is not considered to be a "humane" slaughter technique. A study found that it took ECG patterns between 2 and 5 minutes to decline while the somatosensory responses were present for at least 15 minutes following gill cutting. Despite the ban, exsanguination through gill cutting is still being practiced in countries such as the U.K. and Norway. Exsanguination via decapitation is rarely used in the Netherlands to particularly slaughter eels. Research reveals that eels do not lose visually induced responses for an average of 13 minutes following decapitation. Findings also reveal that eels can remain alive for 8 hours following the cut. "This method is detrimental from an animal welfare perspective" (Bergqvist & Gunnarsson, 2013).

8.5.2.4 Electrocution

If properly executed, electrocution over the head or the entire body, is considered a "more humane" method of stunning. Electrocution could become a threat to welfare when not powerful enough electrical current is applied, leading to fish immobilization without loss of sensibility (Bergqvist & Gunnarsson, 2013).

8.5.2.5 Carbon Dioxide (CO2) Stunning

Carbon dioxide stunning is a commonly used slaughter technique which basically involves the immersion of fish in water with added CO2. Carbon dioxide stunning contributes to "intense avoidance behavior in fish." The cause of death is narcosis. Immobility takes place between 30 seconds and 4 minutes while sensibility remains for up to 5 minutes. This method is not considered humane, yet it is a prevalent finfish aquaculture slaughter method worldwide (Bergqvist & Gunnarsson, 2013).

8.5.2.6 Percussive Stunning

Percussive stunning, "one of the more human methods," if properly executed, involves the application of "a blow to the head using a club." Such method leads to concussion and immediate loss of consciousness or death. In practice, more often than not, several hits are required before loss of consciousness or death, causing "suffering for the fish, since consciousness is not lost immediately" (Bergqvist & Gunnarsson, 2013).

8.5.2.7 Spiking

Spiking, often used on salmon and tuna, is also considered one of the more humane slaughter methods when properly executed. It involves the insertion of a spike in the brain of the fish, leading to immediate loss of consciousness. Stunning results when the spike hits a particular point in the brain. Provided such point is missed, the fish remains with full consciousness and damage to the brain tissue which is "painful." Automatic stunning equipment is recommended as it is considered more accurate (Bergqvist & Gunnarsson, 2013).

8.5.2.8 Salt Bath & Evisceration

Immersion in a salt bath combined with evisceration is one slaughter method used in the Netherlands specifically for eels. In short, eels are inserted in a dry tank, exposed to sodium chloride (NaCl) for approximately 20 minutes, eviscerated and bled until they die. The entire process takes about one hour, a period during which eels may have maintained full consciousness. Eels may remain alive up to 18 hours following the initiation of the procedure. "This is not considered to be a humane method" due to the signs of "severe stress" following the addition of salt to the tank (Bergqvist & Gunnarsson, 2013).

8.5.3 Environment

Aquaculture affects the environment in the following ways:

loss of habitat, transmission of disease to and decline of wild fish populations, waste and chemical emission and invasion of exotic species.

Pesticides, disinfectants, antibiotics, fertilizers and oxidants are used in aquaculture. Water pollution and adverse health impacts on wild aquatic species are two consequences of aquaculture chemical discharge into natural water bodies. Nitrogen (N) leaks out into regional waters in its ammonia and nitrite form and poses toxic risks to fish if in elevated levels. Waste water with surplus feed and faeces can further contaminate regional waters and fish farms. The immune system of aquaculture fish may be compromised as a result of inadequate water quality and crowding conditions.

Natural water habitats for wild fish can be used for aquaculture which can threaten the existence of wild aquatic species and have detrimental environmental consequences such as "loss of nursing grounds and shelter for fish" and sediment transportation. Disease can further spread between wild and farm fish, threatening their continual existence. Farm fish can escape aquaculture and spread diseases to wild fish populations. Farm fish escape can further displace wild fish and thus compromise wild fish biodiversity due to the possibility of genetic superiority of farm fish such as in instances of transgenic fish (Bergqvist & Gunnarsson, 2013).

China's growing economy has contributed to a greater demand for luxury seafood, creating "broader trading networks" that result in "environmental degradation" (Fabinyi et al., 2012).

"The economical and nutritional gains of fish farms are not comparable with the vast environmental and animal welfare losses...In addition, the aquaculture industry has not been able to solve the issue of decline of wild fish populations, but is instead further exacerbating the decline." While "public's interest in environmentally sustainable aquaculture is increasing...ethical consumption is applied only by a small part of the public" (Bergqvist & Gunnarsson, 2013).

8.5.4 Feed

Affluent consumers tend to prefer carnivorous finfish (i.e., salmon, cod, tuna, halibut) and shrimp. The production of carnivorous finfish has been growing by 9% per year since the 1990s. Carnivorous fish species rely on wild fish and fish oil diet. It is therefore unclear as to whether carnivorous fish aquaculture is likely to result "in a net gain, or a net drain, to world fish supplies." The demand for wild fish meal and oil is projected to exceed supply by 2050 provided carnivorous fish aquaculture continues to expand at its current rate. 40% of the world supply of fish meal was allocated to aquaculture in 2002 compared to 10% in 1988 and 22% in 1997. 10 million metric tons (mmt) of wild fish used for feed was consumed by aquaculture in 1997 to produce 20 mmt of farmed fish and shellfish. 17 mmt of wild fish used for feed was used by aquaculture in 2001. Plant-based feed varieties are currently being developed in an attempt to reduce the dependence of wild fish use for fish meal and oil (Naylor & Burke, 2005). Moreover, fossil fuel used for aquaculture feed production and fishing is a major contributor to greenhouse gas (Ziegler et al., 2012).

8.5.5 GM Fish

"The question of the use of GM fish in aquaculture is complex...there is a storm of uncertainties at so many levels" (Coutellec & Doussan, 2012).

8.5.6 Industry & Government

"The lack of protection for...animals is...startling." Shellfish in New Zealand, for example, are on the list of protected by law, yet it is permissible to kill them by boiling them alive and by cutting their limbs and torso while they are still alive

(Zivotofsky, 2012).

Salmon aquaculture has become "highly concentrated." Approximately 30 companies owned two-thirds of the world farmed trout and salmon production in 2001. Marine Harvest, Fjord Seafood, Cermaq and Panfish are the four largest farmed salmon production international companies. The possible lucrative nature of aquaculture "has led many governments to develop policies and programs to support and encourage fish farming." The U.S. Department of Commerce "has a stated goal of increasing the value of the U.S. aquaculture industry from less than $1 billion currently to $5 billion by 2025." Aquaculture companies are "keen on securing a market edge with new products, particularly if they can adopt existing infrastructure and cultivation technology to a broader range of species."

The U.S. federal government "is charged with the management of national resources...in a way that benefits all." The government is to be appropriately compensated for the private use of public resources. Concerns have been expressed that "the aggressive promotion of aquaculture in federal waters" by the U.S. Department of Commerce "will encourage aquaculture practices that benefit only a narrow constituency and that the government (and thus the public) will not be appropriately compensated for the private use of, or harm to, ocean resources" (Naylor & Burke, 2005).

The U.S. Department of State entered into the Pacific Salmon Treaty with Canada in 1999. The purpose of the Treaty was to manage mostly endangered salmon populations originating from the Pacific Northwest and Alaska. The Canadian take levels of salmon are, however, greater than predicted, "jeopardizing the continued existence of the endangered salmon" (Wu, 2009).

Governments "have often promoted aquaculture" as a means of "employment and income generation." In Canada, for example, salmon aquaculture "has been promoted for these reasons." The employment and income benefits from growing aquaculture practices are, however, "controversial." The development of aquaculture and depletion of wild species has contributed to fishermen "losing their jobs" with no guarantee that they "will move into the aquaculture industry." Further, "the employment and income loss in the fish capture industry may be as large, if not larger, than employment and income generation for coastal residents in aquaculture" (Naylor & Burke, 2005).

8.6 Sport Fishing

Recreational sport fishing in the U.S. is a $7 billion industry per year (Turyk et al., 2012).

Discussion

Examples supporting Sosé's theory continue to accumulate. Specific to "Fish Exhaustion" evidence points to wild fish being contaminated with heavy metals such as mercury. The chain of events could be traced back to "Conditions in Motion" with the release of industrial contaminants into the environment. At the end of the chain is the consumer being exposed to heavy metals from consuming contaminated fish. Heavy metal intoxication contributes to adverse health conditions which lead to profit for pharmaceutical and health care systems.

As alluded to earlier, aquatic pollution as well as over-exploitation of fish resources give rise to aquaculture, owned by a small number of companies highly supported by governments. Similarly to water pollution contributing to the growth of the water privatization sector as was observed in "Template for Extinction," pollution and aquatic life extinction leads to the rise of aquaculture. What is public is becoming private. What was once a created problem has now been converted into profit. The deteriorating from aquaculture marine environment does not stop governments and industries from exploiting resources to the maximum in the name of profit.

The above examples not only depict a full circle of how problems (i.e., aquatic pollution; over-fishing) are created to how profit from manufactured problems is generated, but they also provide additional evidence for Sosé's theory that everything in life is interconnected. Creating a problem in one sector (i.e., environmental contamination) generates problems in another (i.e., health). Similarly, profit generated from one sector (i.e., industrial production) leads to profit generated from other (i.e., health care; aquaculture).

In terms of fish welfare, the argument that fish is not a sentient and a conscious being and is therefore exempt from the right to live is undoubtedly inaccurate. Discussed science and fish slaughter methods clearly demonstrate that fish is in fact conscious, can and does feel tremendous amounts of pain during times of mistreatment. It could therefore be argued that the belief that fish is just an object is intentionally propagated (i.e., problem) to increase fish consumption and therefore profit.

An additional example of how problems are converted not only into profit, but also into power concerns fishermen who have been left with little choice, but to work for aquaculture industries, employment which is not even guaranteed. Problems are also translated into profit and power when, similarly to the water privatization sector, humanity is left with no other choice, but to purchase fish from aquaculture industries.

Last, but not least, labeling of fish is highly unreliable, inconsistent and incomplete which implies that truth is being hidden from the public in the name of profit. Humanity would never know what they are consuming which means that they could be consuming anything (as the specific example with the red snapper implies). One such possibility is genetically modified (GM) fish. Demand and thus profit would decrease if humanity is made aware of the fact that they are consuming GM fish which explains why adequate labeling of fish products is literally non-existent.

References

Bergqvist, J. & Gunnarsson, S. (2013). Finfish aquaculture: Animal welfare, the environment, and ethical implications. *Journal of Agricultural & Environmental Ethics, 26,* 75-99.

Binnington, M. J., Quinn, C. L., McLachlan, M. S. & Wania, F. (2014). Evaluating the effectiveness of fish consumption advisories: Modeling prenatal, postnatal, and childhood exposures to persistent organic pollutants. *Environmental Health Perspectives, 112*(2), 178-186.

Bovenkerk, B. & Meijboom, F. L. B. (2013). Fish welfare in aquaculture: Explicating the chain of interactions between science and ethics. *Journal of Agricultural & Environmental Ethics, 26,* 41-61.

Burger, J. (2002). Daily consumption of wild fish and game: Exposures of high end recreationists. *International Journal of Environmental Health Research, 12,* 343-354.

Coutellec, L. & Doussan, I. (2012). Legal and ethical apprehensions regarding relational object. The case of genetically modified fish. *Journal of Agricultural & Environmental Ethics, 25,* 861-875.

Fabinyi, M., Pido, M., Harani, B., Caceres, J., Uyami-Bitara, A., De las Alas, A., Buenconsejo, J. & Ponce de Leon, E. M. (2012). Luxury seafood consumption in China and the intensification of coastal livelihoods in Southeast Asia: The live reef fish for food trade in Balabac, Philippines. *Asia Pacific Viewpoint, 53*(2), 118-132.

Kalshoven, K. & Meijboom, F. L. B. (2013). Sustainability at the crossroads of fish consumption and production ethical dilemmas of fish buyers at retail organizations in The Netherlands. *Journal of Agricultural & Environmental Ethics, 26,* 101-117.

Kissinger, L., Lorenzana, R., Mittl, B., Lasrado, M., Iwenofu, S.,

Olivo, V., Helba, C., Capoeman, P. & Williams, A. H. (2010). Development of a computer-assisted personal interview software system for collection of tribal fish consumption data. *Risk Analysis, 30*(12), 1833-1841.

Madhoo, Y. N. (2011). Fish imports as an environmental policy. *Applied Economics Letters, 18*, 859-864.

Mitchell, M. (2011). Increasing fish consumption for better health – are we being advised to eat more of an inherently unsustainable protein? *Nutrition Bulletin, 36*, 438-442.

Naylor, R. & Burke, M. (2005). Aquaculture and ocean resources: Raising tigers of the sea. *Annual Review of Environment & Resources, 30*, 185-218.

Oken, E., Choi, A. L., Karagas, M. R., Marien, K., Rheinberger, C. M., Schoeny, R., Sunderland, E. & Korrick, S. (2012). Which fish should I eat? Perspectives influencing fish consumption choices. *Environmental Health Perspectives, 120*(6), 790-798.

Pastorelli, A. A., Baldini, M., Stacchini, P., Baldini, G., Morelli, S., Sagratella, E., Zaza, S. & Ciardullo, S. (2012). Human exposure to lead, cadmium and mercury through fish and seafood product consumption in Italy: A pilot evaluation. *Food Additives & Contaminants: Part A, 29*(12), 1913-1921.

Tortfadottir, J. E., Valdimarsdottir, U. A., Mucci, L. A., Kasperzyk, J. L., Fall, K., Tryggvadottir, L., Aspelund, T., Olafsson, O., Harris, T. B., Jonsson, E., Tulinius, H., Gudnason, V., Adami, H.-O., Stamfer, M. &Steingrimsdottir, L. (2013). Consumption of fish products across the lifespan and prostate cancer risk. *PLoS ONE, 8*(4), 1-11.

Turyk, M. E., Bhavsar, S. P., Bowerman, W., Boysen, E., Clark, M., Diamond, M., Mergler, D., Pantazopoulos, P., Schantz, S. & Carpenter, D. O. (2012). Risks and benefits of consumption of Great Lakes fish. *Environmental Health Perspectives, 120*(1), 11-18.

Wu, F. & Guclu, H. (2013). Global maize trade and food security: Implications from a social network model. *Risk Analysis,*

33(12), 2168-2178.

Ziegler, F., Winther, U., Hognes, E. S., Emanuelsson, A., Sund, V. & Ellingsen, H. (2012). The carbon footprint of Norwegian seafood products on the global seafood market. *Journal of Industrial Ecology, 17*(1), 103-116.

Zivotofsky, A. Z. (2012). Government regulations of *Shechita* (Jewish religious slaughter) in the twenty-first century: Are they ethical? *Journal of Agricultural & Environmental Ethics, 25,* 747-763.

Chapter 9

DEFORMATION

Overview

"Deformation" describes erosion types, causes, processes and projections, presents concrete examples of erosion and proposes solutions to preventing erosion.

9.1 Soil Erosion

Soil erosion is defined as "a geophysical process involving the detachment, transport and deposition of soil materials." Soil erosion is one of the most significant ecological challenges in many parts of Earth, putting "great pressures on the earth surface and its environment." Soil erosion has been characterized by scholars as "earth cancer." It is a common form of land degradation and is "responsible for enormous losses in soil fertility and degradation of water resources." Research findings reveal that more than 85% of land degradation worldwide is correlated to soil erosion. A major portion of soil erosion took place after World War II, contributing to a 17% decline in crop productivity and environmental deterioration. Changes to landform could contribute to a decrease in rainfall infiltration rates and local slope stability, as well as an acceleration of soil erosion (Wei et al, 2009).

Two main factors shape the structural crust of soil: the impact from the kinetic energy of raindrops and the degree of stability of soil composites. Soil erosion, runoff and ponding are formed when the hydraulic conductivity of the structural crust is less than the magnitude of the rainfall. The structural crust is formed within several minutes, decreasing the infiltration rate of the soil, thus posing great concern for cultivated and uncultivated land (Goldshleger et al., 2009).

9.2 Water Erosion

Water and wind erosion are the two major types of erosion. "Water erosion is the most destructive erosion type worldwide," contributing to land and environmental degradation including soil erosion. Water erosion contributes to 55% of total global erosion. Water erosion has greater land and environmental impact compared to wind erosion. Water erosion further contributes to atmospheric carbon dioxide (CO_2) emission thus being "partly responsible" for the intensified greenhouse effect. Preventing water erosion is thereafter "of paramount importance in the management and conservation of natural resources."

Precipitation is the primary factor shaping erosion variation (Wei et al., 2009). Soil conservation is proposed as a solution (Brazier, 2004). Rainfall intensity and amount resulting from climate change is projected to continue increasing worldwide, possibly intensifying water erosion (Wei et al., 2009).

9.3 Wind Erosion

"Wind erosion is one of the most serious environmental problems in the arid, semi-arid and dry subhumid areas around the world." Wind erosion contributes to substantial crop damages and adverse health effects resulting from dust storms. The United States (U.S.) and Canada are the two countries subjected to wind erosion to the greatest degree (Shi et al., 2004).

9.4 Mountain Erosion

A recent study demonstrate that worldwide rates of mountain erosion have increased in the past 6 million years and most extremely, since the last 2 million years (Herman et al., 2013).

9.5 Coastal Erosion

9.5.1 Quintana Roo, Mexico

"Shore erosion is a serious problem" prevalent in a number of countries with coastal areas. The coasts of the Caribbean Sea, the Gulf of Mexico and the Pacific Ocean are affected by erosion. The coastal regions of the state of Quintana Roo in Mexico have suffered from erosion, particularly in the past decade. Storms and hurricanes with higher than usual intensity, most likely due to global warming and anthropogenic activities, such as the construction of hotels have contributed to the increasing coastal erosion in Quintana Roo (Lopez, 2014).

9.5.2 Latin America

It is common and substantial in some areas of coastal Latin America. Erosion often occurs as a result of anthropogenic events. Permanent erosion results from deforestation, fragmentation of the coast, land use alterations, construction of infrastructure and tectonic subsidence. Climate change and the associated sea level rise (SLR) may contribute to additional coastal erosion at the coastal regions of Latin America (Silva et al., 2014).

9.5.3 Riviera Maya, Mexico

The region between Punta Bete and Punta Maroma on the Riviera Maya, Mexico suffers from chronic erosion due to anthropogenic (i.e., shoreline constructions compromising the transport of sediment) causes (Oderiz et al., 2014).

9.5.4 Balneario Solfs, Uruguay

Balneario Solfs is a coastal resort located in Uruguay. Erosion has been affecting the main beach of the resort, contributing to 35 m of cliff recession between 1980 and 2001, loss of sandy beach area, road damage, affected tourism (the major economic asset) and a risk of collapsing buildings. The main anthropic causes of erosion are possibly the construction of structures that interfere with littoral transportation as well as river damming leading to reduced sediment supply. Beach nourishment has been proposed as an effective methodology to restore the integrity of the beach area (Alonso et al., 2014).

9.5.5 Baia de Guanabara, Brazil

Rising urban and industrial development associated to oil and gas industry have contributed to coastal erosion at Baia de Guanabara which is one of the largest bay areas of Brazil (Araruna Junior et al., 2014).

9.6 Beach Erosion

Sandy beaches are located along most of the oceans of the world. They hold economic, social (i.e., recreation) and ecological (i.e., filtering seawater; recycling nutrients) value. Sandy beaches are habitat to "distinctive biodiversity" such as hundreds of invertebrate species. Yet, rising sea levels, increasing human use and coastal infrastructure are all contributing to accelerating anthropogenic pressures. Sandy beaches are "increasingly becoming trapped in a 'coastal squeeze' between burgeoning human populations from the land and the effects of global climate change from the sea." Methodologies, such as beach nourishment and shoreline armoring designed to counteract alterations of beach systems (erosion and shoreline retreat), "can result in severe ecological impacts and loss of biodiversity at local scales, but are predicted also to have cumulative large-scale consequences worldwide." Direct conservation efforts are required to offset the escalating deterioration of sandy beaches (Schlacher et al., 2007).

9.6.1 Pau Amarelo Beach, Brazil

Urbanization coupled with soil use and meteo-oceanographic processes have contributed to increased rates of erosion of Brazilian beach areas. Beaches are of great importance as recreational regions. Beaches also serve as a "natural defense system for the coast" (Gomes & da Silva, 2014).

Pau Amarelo Beach in Pernambuco, Brazil, has experienced erosion. The beach has a significant tourist, community and local economic value. Anthropogenic (i.e., poor regional planning compromising the natural balance of sediments) factors are believed to contribute to Pau Amarelo Beach erosion. There are a number of strategies that can be used to protect beach areas from erosion. Methods to prevent beach erosion include construction of hard structures such as seawalls and groins as well as utilization of soft technologies such as artificial nourishment and sand-by-passing. Beach nourishment which involves the addition of sand to widen the eroded beach area, is the most commonly used soft methodology to protect and restore the integrity of the beach and the associated environment. Beach nourishment is suggested as an optimal solution to Pau Amarelo Beach erosion as "it works with nature and not against it" (Martins & Pereira, 2014).

9.6.2 Matanchen Bay, Mexico

Matanchen Bay, located on the Pacific coast of Nayarit state, Mexico, is a part of a natural conservation area. Erosion processes have been observed at Matanchen Bay. Anthropogenic activities, such as the construction of a dam system, a transport infrastructure, a harbor and tourist facilities in the bay and surrounding areas are the primary causes of erosion. Beach erosion at Matanchen Bay is projected to continue if no action is undertaken. Supplying the system with artificial sediment is recommended as a means to counteract the land erosion that may take place (Martinez et al., 2014).

9.6.3 Isla del Carmen Island, Gulf of Mexico

Isla del Carmen Barrier Island on the Gulf of Mexico is affected by beach erosion due to human activities such as construction, artificial lagoon openings and damages to vegetation (Escudero et al., 2014).

9.6.4 Sea Turtles

Sea turtles in the U.S. are an endangered species. Shoreline erosion contributes to the "loss" of sea turtles "nesting habitat". In certain areas, beach erosion and storm events contribute to a 60% annual loss of sea turtle nests laid in the U.S. It is demonstrated that recycled glass cullets which are similar to natural sand, could be successfully used for beach fill and as sea turtle nesting substrate. Recycled glass cullets could represent a "new methodology of beach protection" (Makowski et al., 2008).

9.7 Lagoon Erosion

9.7.1 Barra de Navidad, Mexico

Anthropogenic activities (i.e., poor planning; removal of sandspit; construction of breakwater and tourist infrastructure; cutting of extensive areas of mangrove) in the past 4 decades have contributed to detrimental changes to Barra de Navidad lagoon system in Mexico such as a substantial retreat of coastline area (Gonzalez-Vazquez et al.,

2014).

9.7.2 Venice Lagoon, Italy

The Venice Lagoon is an aquatic lagoon ecosystem located in Italy. Anthropogenic (i.e., industrial construction) activities have contributed to an increase in erosion rate to an annual net loss of 800,000 m³ of sediment (Bellucci et al., 2013).

9.8 Barrier-island Chains

"Barrier-island chains worldwide are undergoing substantial changes, and their futures remain uncertain." The Mississippi barriers in the Gulf of Mexico are suffering accelerated land loss and translocation. The Dauphin Island is also experiencing land loss. The major reasons for land loss are frequent and intense storm events, rising of the sea level and deficit in sediment. Based on expected trends in SLR and storms resulting from global warming, it is projected that the Mississippi-Alabama (MS-AL) barrier islands will continue to experience accelerated rates of land loss (Morton, 2008).

9.9 Jamaica Bay, New York

"The marsh islands of Jamaica Bay, New York, are disappearing." The interior marshes of Jamaica Bay are projected to perish by 2024. Jamaica Bay is 107 km² or 26,640 acres. In spite of its deteriorated water quality, it is a habitat to marine and avian species. 91 fish and 325 bird species have been identified at Jamaica Bay. It is a recreational destination for 8 million New York City residents affected by SLR, loss of sediment, anthropologic physical changes and pollution. The surface area of Jamaica Bay has decreased from 101 km² (25,000 acres) in the mid nineteenth century to 53 km² (13,000 acres) at present time. The volume of the bay has however increased by 350%. "Inundation, including sea level rise" results in increase in interior marsh island tidal pools, decline in marsh grass density, deterioration of the peat root system and an increase in the potential for erosion. "Some 905 x 10 6 L/d or 239 x 10 6 gal/d of sewage affluent are discharged into Jamaica Bay" (Swanson & Wilson, 2008).

Discussion

World War II contributed to major soil erosion on a global level, resulting in 17% decrease in soil productivity. As will be explored in "Salvation," the Green Revolution took place right after World War II as a means to increase food production and to alleviate hunger. Genetically modified (GM) crops were advertised as the only alternative to increasing food production. It could be deduced that World War II was the created problem which contributed to soil erosion and served as the basis for generating profit from GM crops.

References

Alonso, R., Lopez, G., Mosquera, R., Solari, S. & Teixeira, L. (2014). Coastal erosion in Balneario Solís, Uruguay. *Journal of Coastal Research, 71*, 48-54.

Araruna Junior, J. T., Pereira de Campos, T. M., Pires, P. J. M. (2014). Sediment characteristics of an impacted coastal bay: Baía de Guanabara, Rio de Janeiro, Brazil. *Journal of Coastal Research, 71*, 41-47.

Bellucci, L. G., Mugnai, C., Guiliani, S., Romano, S., Albertazzi, S. & Frignani, M. (2013). PCDD/F contamination of the Venice Lagoon: A history of industrial activities and past management choices. *Aquatic Ecosystem Health & Management, 16*(1), 62-69.

Brazier, R. (2004). Quantifying soil erosion by water in the UK: A review of monitoring and modelling approaches. *Progress in Physical Geography, 28*(3), 340-365.

Escudero, M., Silva, R. & Mendoza, E. (2014). Beach erosion

driven by natural and human activity at Isla del Carmen Barrier Island, Mexico. *Journal of Coastal Research, 71*, 62-74.

Goldshleger, N., Ben-Dor, E., Chudnovsky, A. & Agassi, M. (2009). Soil reflectance as a generic tool for assessing infiltration rate induced by structural crust for heterogeneous soils. *European Journal of Soil Science, 60*, 1038-1051.

Gomes, G. & da Silva, A. C. (2014). Coastal erosion case at Candeias Beach (NE-Brazil). *Journal of Coastal Research, 71*, 30-40.

Gonzalez-Vazquez, J. A., Silva, R., Mendoza, E., Delgadillo-Calzadilla, M. A. (2014). Towards coastal management of a degraded system: Barra de Navidad, Jalisco, Mexico. *Journal of Coastal Research, 71*, 107-113.

Herman, F., Seward, D., Valla, P.G., Carter, A., Kohn, B., Willett, S. & Ehlers, T. A. (2013). Worldwide acceleration of mountain erosion under a cooling climate. *Nature, 504*, 423-436.

Lopez, R. (2014). Beach restoration at Grand Velas Hotel, Riviera Maya, Mexico. *Journal of Coastal Research, 71*, 86-92.

Makowski, C., Rusenko, K. & Kruempel, C. J. (2008). Abiotic suitability of recycled glass cullet as an alternative sea turtle nesting substrate. *Journal of Coastal Research, 24*(3), 771-779.

Martinez, M. R. E., Silva, R. & Mendoza, E. (2014). Identification of coastal erosion causes in Matanchén Bay, San Blas, Nayarit, Mexico. *Journal of Coastal Research, 71*, 93-99.

Martins, K. A. & Pereira, P. S. (2014). Coastal erosion at Pau Amarelo Beach, northeast of Brazil. *Journal of Coastal Research, 71*, 17-23.

Morton, R. (2008). Historical changes in the Mississippi-Alabama barrier-island chain and the roles of extreme storms, sea level, and human activities. *Journal of Coastal Research, 24*(6), 1587-1600.

Oderiz, I., Mendoza, E., Leo, C., Santoyo, G., Silva, R., Martinez,

R., Grey, E. & Lopez, R. (2014). An alternative solution to erosion problems at Punta Bete-Punta Maroma, Quintana Roo, Mexico: Conciliating tourism and nature. *Journal of Coastal Research, 71,* 75-85.

Schlacher, T. A., Dugan, J., Schoeman, D. S., Lastra, M., Jones, A., Scapini, F., McLachlan, A. & Defeo, O. (2007). Sandy beaches at the brink. *Diversity and Distributions, 13,* 556-560.

Shi, P., Yan, P. Yuan, Y. & Nearing, M. A. (2004). Wind erosion research in China: Past, present and future. *Progress in Physical Geography, 28*(3), 366-386.

Silva, R., Martinez, M. L., Hesp, P. A., Catalan, P., Osorio, A. F., Martell, R., Fossati, M., da Silva, G. M., Marino-Tapia, I., Pereira, P., Cienguegos, R., Klein, A. & Govaere, G. (2014). Present and future challenges of coastal erosion in Latin America. *Journal of Coastal Research, 71,* 1-16.

Swanson, R. L. & Wilson, R. E. (2008). Increased tidal ranges coinciding with Jamaica Bay development contribute to marsh flooding. *Journal of Coastal Research, 24*(6), 1565-1569.

Wei, W., Chen, L. & Fu, B. (2009). Effects of rainfall change on water erosion processes in terrestrial ecosystems: A review. *Progress in Physical Geography, 33*(3), 307-318.

Chapter 10

SUFFOCATION

Overview

"Suffocation" concerns air pollution (including indoor air pollution), its causes and consequences on human and environmental health. Solutions to air pollution are introduced.

***

Atmospheric pollution remains "a major environmental and public health issue" (Deguen et al., 2012). The combustion of fossil fuels is the primary contributor to ambient air pollution (Bergin et al., 2005).

10.1 Air Pollution

Acids, particulate matter (PM), ozone (O3), mercury (Hg) and persistent organics (POPs) are significant air pollutants that originate from anthropogenic processes. All air pollutants enlisted above have been found to cause reproductive, nervous, immune and behavioral disorders while also having a tendency to remain in the environment for many years (Bergin et al., 2005). The developing fetus, children, the elderly, people with diabetes, hypertension, pre-existing cardiovascular and respiratory disease are most susceptible to the adverse health effects of atmospheric pollution (Peel et al., 2013).

One quarter of the world's population is estimated to be exposed to air pollutants. Approximately 6.4 million years of healthy life is being lost due to long-term exposure to atmospheric PM alone (Bergin et al., 2005). Acute exposure to gaseous air pollution is correlated with increased mortality and morbidity from respiratory and cardiac disease (Cakmak et al., 2007). Present levels of ambient air pollution are associated with cardiac and respiratory morbidity and mortality in the general population in the Americas, Australia and Europe (Dales et al., 2006). Atmospheric pollution contributes to one in eight deaths or 2.5 million deaths per annum (Neira, 2014). About half of the American population resides in regions characterized by unhealthy levels of air pollution. Over 70,000 people die from air pollution in the United States (U.S.) per annum (Ozymy & Jarrell, 2011).

Researchers found that PM 2.5 and O3 levels in the U.S. in 2005 "were responsible" for 130,000 to 320,000 PM 2.5 and 4,700 O3-related premature deaths which amounts to 1 in 20 or 6.1% of total death in the U.S. Exposure to PM 2.5 and O3 in 2005 in the U.S. further contributed to close to 200,000 non-fatal heart attacks and 90,000 hospital admissions resulting from cardiovascular or respiratory illness. The adverse health effects resulting from air pollution in the U.S. "also pose a significant economic burden, through hospital admissions, emergency room visits, and lost work and school days" (Peel et al., 2013). The monetary cost relative to the health effects of ambient air pollution is "extremely high" (Marzouk & Madany, 2012).

10.2 Air Pollution & Children

Exposure to air pollution has been associated with childhood asthma, cancer, respiratory illness and heart disease (Ozymy & Jarrell, 2011). Exposure to PM 2.5 and O3 in 2005 in the U.S. contributed to 2.5 million cases of pediatric asthma (Peel et al., 2013). Children's immune system and lungs are developing and exposure to high concentrations of ambient O3 may interfere with normal developmental processes (Lin et al., 2008b). It is demonstrated that present levels of gaseous air pollution in Canada are correlated with "significant proportion of hospitalization for respiratory disorders in neonates" (Dales et al., 2006).

Reduced birth weight has been linked to cognitive developmental delays, greater risk of type II diabetes, hypertension and coronary heart disease during adulthood (Choi et al., 2006). 20 million infants worldwide are affected by low birth weight (LBW) (Kannan et al., 2006). Study findings reveal that maternal exposure to ambient pollution could increase the risk of LBW even in low concentrations (Bell et al., 2007).

10.3 Urban Air Pollution

"Air pollution is one of the top environmental concerns for citizens." 3.3 billion people worldwide live in cities (Minguez et al., 2013).

10.4 Air Pollution, Class and Ethnicity

Ambient pollution contributes to adverse health consequences in developed and developing countries. Socially disadvantaged people and individuals with poor health are affected by air pollution to the largest extent. Globalization has contributed to the shifting of industries "notorious for their pollution" from developed to developing areas where "costs of production are cheaper and environmental regulations are less stringent" (Wong et al., 2008). Atmospheric pollution has a large impact on urban areas in developing countries (Deguen et al., 2012).

White and Hispanic populations in the U.S. tend to live "segregated from one another." Areas with predominantly minority populations tend to be situated in greater proximity to air pollution sources such as roadway traffic. One study demonstrated that Hispanic neighborhoods in six U.S. cities were exposed to higher concentrations of air pollution compared to primarily White neighborhoods (Jones et al., 2014).

10.5 Indoor Air Pollution (IAP)

About one half of the global population relies on biomass (i.e., wood) and coal for heating and cooking (McCracken et al., 2007). Indoor air pollution (IAP) from fuel combustion used for lighting, cooking and heating is a major risk factor for morbidity and mortality (Gall et al., 2013). A study revealed that 1.6 million premature deaths and 3.6% of the worldwide prevalence of disease result from solid fuel use indoors (McCracken et al., 2007). IAP caused nearly 2 million deaths in 2001 and 3% of the "global burden of disease." Respiratory diseases such as acute respiratory infections (ARIs), LBW, cataracts, lung cancer, perinatal and infant mortality result from indoor solid fuel use. IAP has a disproportionate impact on children and women in developing areas (Gall et al., 2013).

10.6 Wildfires & Prescribed Fires

Wildfires generate smoke that is primarily composed of carbonaceous aerosol. Some wildfires contain organic compounds, known mutagens, irritants and carcinogens. Exposure to wood smoke contributes to impaired respiratory function, mortality, respiratory symptoms and worsening of asthma (Mauderly & Chow, 2008). Exposure to wildfire smoke is correlated to increased visits in emergency departments, particularly among human beings with pre-existing cardiopulmonary conditions. Increased frequency and intensity of fires is projected to take place as a result of higher temperatures, prolonged drought periods and decreased soil moisture (Peel et al., 2013).

10.7 Regulation

The Clean Air Act (CAA) is responsible for controlling atmospheric pollution in the U.S. (Peel et al., 2013). Section 112 of the CAA regulates the health risks associated with exposure to hazardous air pollutants (HAPs). HAPs are defined as atmospheric pollutants which "present, or may present...a threat of adverse human health effects...or adverse environmental effects." HAPs are categorized as carcinogenic (which cause a variety of cancers) and non-carcinogenic (which cause reproductive, neurobiological or acute illnesses) (Jackson, 2009). The CAA has identified 188 HAPs (Ozymy & Jarrell, 2011). Section 112 of the CAA is implemented by the EPA. EPA managed to establish standards only for 8 HAPs as a means of avoiding "unduly costly regulations" (Jackson, 2009). U.S. state and federal efforts to attain air quality standards have been largely unsuccessful (Burtraw et al., 2005).

"Command and control environmental regulations in the United States have typically sought to control pollution after it has been generated." The National Pollution Prevention Act enacted in 1990 recommends prevention or reduction of pollution from its source when possible, yet the Act does not make pollution prevention strategies mandatory (Harrington et al., 2014).

10.8 Transboundary Air Pollution

Acids, PM and O3 and to a lesser degree Hg and POPs "are tightly linked chemically and are, indeed, virtually inseparable." Such pollutants are capable of being transported from 100 to a couple of thousands of kilometers and are therefore able to cross territorial boundaries. Air pollution in one jurisdiction (i.e., state; province; nation) may therefore be caused by pollution emission from another jurisdiction. Overcoming cultural, political and economic differences and establishing grounds for cooperation among different jurisdictions is imperative if ambient air pollution is to be managed effectively. While most national governments regulate air pollution via the establishment of air quality standards, international governmental authority pertaining to air pollution is often times lacking (Bergin et al., 2005).

Loadings of dust from the Saharan-Sahel region to the

atmosphere is about 1 billion tons per year. Satellite images reveal that dust originating from Africa is transported to Europe across the Mediterranean Sea and subsequently crosses the Atlantic Ocean, affecting distant regions. Dust flux affects visibility, climate and human health. According to the WHO, dust and drought in Africa's sub-Saharan region caused meningococcal meningitis, contributing to 250,000 cases and 25,000 deaths in 2006. Approximately one-half of the dust mass is composed of particles less than 2.5 mm (Polymenakou et al., 2008).

Polymenakou et al. (2008) revealed that a number of small particle size pathogens are present in Saharan dust storm with the potential for transboundary atmospheric transportation, adverse health, ecosystem and agricultural impacts.

10.9 Air Quality

Programs and technologies designed to improve ambient quality are "expected to cost trillions of dollars worldwide" (Shin et al., 2008). Air pollution is projected to continue rising as the global population and fossil fuel combustion increase (Bergin et al., 2005).

Discussion

Research demonstrates that air pollutants cause behavioral disorders. One can only imagine the extent of profit generated by the mental health, education and judicial sectors (as could be seen in "Noun=Verb") from behavioral disorders. The scope of profit acquired from physiological childhood illness is also within the reach of imagination. It is apparent how air pollution is the orchestrated problem from which governmental and industrial profit is acquired.

Similarly to the U.S. Marine Mammal Protection Act and the EPA, the Clean Air Act (CAA) is only in operation as a means to deceive the public that effective measures are taken to intervene on the level of air pollution when in actuality this is not so. The root of the problem or fossil fuel combustion, remains unattended to while the acquisition of profit continues from the creation of atmospheric problems. Unaware public translates into uninterrupted financial flow.

References

Bell, M. L., Dominici, F., Ebisu, K., Zeger, S. L. & Samet, J. M. (2007). Spatial and temporal variation in PM2.5 chemical composition in the United States for health effects studies. *Environmental Health Perspectives, 115*(7), 989-995.

Bergin, M. S., West, J. J., Keating, T. J. & Russell, A. G. (2005). Regional atmospheric pollution and trasnboundary air quality management. *Annual Review of Environment & Resources, 30*, 1-37.

Burtraw, D., Evans, D. A., Krupnick, A., Palmer, K. & Toth, R. (2005). Economics of pollution trading for SO2 and NOX. *Annual Review of Environment & Resources, 30*, 253-289.

Cakmak, S., Dales, R. E. & Vidal, C. B. (2007). Air pollution and mortality in Chile: Susceptibility among the elderly. *Environmental Health Perspectives, 115*(4), 524-527.

Choi, H., Jedrychowski, W., Spengler, J., Camann, D. E., Whyatt, R. M., Rauh, V., Tsai, W.-Y. & Perera, F. P. (2006). International studies of prenatal exposure to polycyclic aromatic hydrocarbons and fetal growth. *Environmental Health Perspectives, 114*(11), 1744-1750.

Dales, R. E., Cakmak, S. & Doiron, M. S. (2006). Gaseous air pollutants and hospitalization for respiratory disease in the neonatal period. *Environmental Health Perspectives, 114*(11), 1751-1754.

Deguen, S., Segala, C., Pedrono, G. & Mesbah, M. (2012). A new air quality perception scale for global assessment of air pollution health effects. *Risk Analysis, 32*(12), 2043-2054.

Gall, E. T., Carter, E. M., Earnest, C. M. & Stephens, B. (2013).

Indoor air pollution in developing countries: Research and implementation needs for improvements in global public health. *American Journal of Public Health, 103*(4), e1-e6.

Harrington, D. R., Deltas, G. & Khanna, M. (2014). Does pollution prevention reduce toxic releases? A dynamic panel data model. *Land Economics, 90*(2), 199-221.

Jackson, A. (2009). EPA's fuzzy bright line approach to residual risk. *Ecology Law Quarterly, 36,* 439-466.

Jones, M. R., Diez-Roux, A. V., Hajat, A., Kershaw, K. N., O'Neill, M. S., Guallar, E., Post, W. S., Kaufman, J. D. & Navas-Acien, A. (2014). Race/ethnicity, residential segregation, and exposure to ambient air pollution: The Multi-Ethnic Study of Atherosclerosis (MESA). *American Journal of Public Health, 104*(11), 2130-2137.

Kannan, S., Misra, D. P., Dvonch, T. & Krishnakumar, A. (2006). Exposures to airborne particulate matter and adverse perinatal outcomes: A biologically plausible mechanistic framework for exploring potential effect modification by nutrition. *Environmental Health Perspectives, 114*(11), 1636-1642.

Lin, S., Liu, X., Le, L. H. & Hwang, S.-A. (2008b). Chronic exposure to ambient ozone and asthma hospital admissions among children. *Environmental Health Perspectives, 116*(2), 1725-1730.

Marzouk, M. & Madany, M. (2012). Health effects associated with passenger vehicles: Monetary values of air pollution. *Archives of Environmental & Occupational Health, 67*(3), 145-154.

Mauderly, J. L. & Chow, J. C. (2008). Health Effects of Organic Aerosols. *Inhalation Toxicology, 20,* 257-288.

McCracken, J. P., Smith, K. R., Diaz, A., Mittleman, M. A. & Schwartz, J. (2007). Chimney stove intervention to reduce long-term wood smoke exposure lowers blood pressure among Guatemalan women. *Environmental Health Perspectives, 115*(7), 996-1001.

Minguez, R., Montero, J.-M. & Fernandez-Aviles, G. (2013).

Measuring the impact of pollution on property prices in Madrid: Objective versus subjective pollution indicators in spatial models. *Journal of Geographical Systems*, *15*, 169-191.

Neira, M. (2014). The 2014 WHO conference on health and climate. *Bulletin of the World Health Organization*, *92*, 546.

Ozymy, J. & Jarrell, M. L. (2011). Upset over air pollution: Analyzing upset event emissions at petroleum refineries. *Review of Policy Research*, *28*(4), 365-381.

Peel, J. L., Haeuber, R., Garcia, V., Russell, A. G. & Neas, L. (2013). Impact of nitrogen and climate change interactions on ambient air pollution and human health. *Biogeochemistry*, *114*, 121-134.

Polymenakou, P. N., Mandalakis, M., Stephanou, E. G., Tselepides, A. (2008). Particle size distribution of airborne microorganisms and pathogens during an intense African dust event in the Eastern Mediterranean. *Environmental Health Perspectives*, *116*(3), 292-296.

Shin, H. H., Stieb, D. M., Jessiman, B., Goldberg, M. S., Brion, O., Brook, J., Ramsay, T. & Burnett, R. T. (2008). A temporal, multicity model to estimate the effects of short-term exposure to ambient air pollution on health. *Environmental Health Perspectives*, *116*(9), 1147-1153.

Wong, C.-M., Ou, C.-Q., Chan, K.-P., Chau, Y.-K., Thach, T.-Q., Yang, L., Chung, R. Y.-N., Thomas, G. N., Peiris, J. S. M., Wong, T.-W., Hedley, A. J. & Lam, T.-H. (2008). The effects of air pollution on mortality in socially deprived urban areas in Hong Kong, China. *Environmental Health Perspectives*, *116*(9), 1189-1194.

Chapter 11

PARADOXICAL FORMATION

Overview

"Paradoxical Formations" explores global warming and climate change. Specifically, it delves into topics such as greenhouse gas, runaway greenhouse, resulting from climate change health and environmental effects (i.e., migration of species; sea level rise). "Paradoxical Formations" concludes by proposing solutions to climate change.

11.1 Global Warming

Anthropogenic greenhouse gas (GHG) emissions are contributing to global warming (Ferraro et al., 2014). Significant anthropogenic global warming has taken place in the past 5 decades (Sang, 2013). The global mean surface air temperature (SAT) has increased by 0.75 degrees C^0 in the past century and by 0.2 degrees C^0 /decade since 1971. The average temperature of the oceans' surface layers has increased by 0.6 degrees C^0 between 1900s and 2010 (Paeth et al, 2009).

A global warming of 4.5 degrees C^0 is projected to occur by the end of the 21st century in case that appropriate measures to counteract atmospheric GHG emissions are not taken (Dilling & Hauser, 2013).

11.2 Greenhouse Gas (GHG)

Historically, only developed countries had established policies geared towards the reduction of GHG emissions. However, such situation is expected to shift as the rate of GHG emission in developing countries is projected to surpass that of developed countries by the year 2020. Developed and developing countries would be financially penalized by the Kyoto Protocol for not succeeding in meeting GHG level targets (Dhansay et al., 2014). The Kyoto Accord was established in response to global climate change and as a means to reduce GHG emissions. The United States (U.S.) has refused to become a signatory of the Kyoto Accord (Strachan et al., 2006). The South African government established a commitment to reduce GHG emission by 40% by 2050. The rising demand for energy, however, lead to the commissioning of two additional coal power plants which are expected to increase GHG emissions even further. Alternative energy sources are being explored by the South African government. One such proposal is the enhanced geothermal system (EGS) which consists of hydraulic fracturing and the generation of organic steam used to produce electricity. EGS is a renewable energy source that is costly and is not currently commercially deployed in South Africa, yet a provision is in place for its inclusion (Dhansay et al., 2014).

Results from previous studies indicate that household consumption is responsible for 13% - 35% of a country's total and direct emission of GHG. Household consumption contributes between 60% and 80% of a country's GHG emission when indirect emissions are included (Wilson et al., 2013).

11.3 Runaway Greenhouse

A "runaway greenhouse" results from the formation of a hot atmosphere rich in water vapour that prevents the thermal radiation emission into space. Such climatic emergency would contribute to an evaporation of the water on earth and an extinction of all life on the planet. The greater the warming of the planet, the greater the evaporation of water from earth. Water vapor is a greenhouse gas which increases the warming of earth. In short, as surface temperatures increase, the atmosphere's thickness increases as a result of water vapour, precluding the amount of thermal radiation emitted in space. Runaway warming results from the planet absorbing more solar energy than it could emit back in space. Runaway greenhouse occurred in Venus. Earth is projected to undergo runaway greenhouse in approximately 2 billion years as the sun's luminosity continues to increase. Almost all current evidence points to the fact that runaway greenhouse would not occur prematurely as a result of humanity's "climate-altering activities." However, the complexity of the climate system on earth prevents scientists from entirely ruling out "the possibility that human actions might cause a transition, if not to full runaway, then at least to a much warmer climate state than the present one" (Goldblatt & Watson, 2012).

International organizations have not been successful in encouraging global efforts towards mitigation of GHG emissions. Diverging national interests and high cost for unclear benefits hinder mitigation efforts (Victor, 2008). Though historically the U.S. "has stood in the way of international progress on this issue," President Obama initiated "a global effort toward a mitigation agreement" (Robock et al., 2009). Mitigation efforts require participation on a global scale (Victor, 2008).

11.4 Climate Change

Climate change is occurring on a global level (Paeth et al., 2009). Earth's climate is being changed as a result of anthropogenic GHG emissions (Goes et al., 2011). "More extremes in the climate are already visible in European weather observations" (van Wezel et al., 2006). Precipitation is one of the most significant climatic factor for understanding climatic systems (Sang, 2013). Climate change has an impact on the intensity and frequency of precipitation (Pendergrass & Hartmann, 2014). Climate change and human activities are altering rainfall patterns which are projected to "become more pronounced in future decades." The total amount of precipitation increased in the U.S. during the twentieth century. 53% of such increase occurred in the form of extreme precipitation events (Wei et al., 2009). The southwestern area of the U.S. is projected to become drier with more intense precipitation events. The western-north parts of the U.S. have been subject to a 15% decrease in precipitation since 1900 (Barron-Gafford et al., 2013). It is predicted that "some present-day climates may disappear" (Paeth et al., 2009).

11.5 Carbon Dioxide (CO2)

Carbon dioxide (CO2) is the most significant anthropogenic greenhouse gas. Energy consumption, fossil fuel combustion, cement manufacture and changes in land use are major sources of CO2 emission in the atmosphere (Hou et al., 2013). Mean global atmospheric CO2 levels are increasing as a result of growing CO2 emissions from anthropogenic activities (Crous et al., 2010). Atmospheric CO2 is the leading cause of climate change. CO2 concentrations "are now higher than any experienced in the last 800,000 years" (Paeth et al., 2009). Researchers suggest that approximately 40% more CO2 is present in the atmosphere in comparison to levels prior to the Industrial Revolution (Farahani, 2014). The rate of CO2 emission in the atmosphere is rising (Robock et al., 2009). The rising of global atmospheric CO2 concentrations is estimated to contribute to more extreme weather events and rising sea levels (Hou et al., 2013).

"At present, the Earth's vegetation is exposed to atmospheric CO2 concentrations that have been unprecedented over the past 15 million years." Land ecosystems absorb 25-30% of anthropogenic CO2 emissions (Bader et al., 2013). Nitrogen (N) takes part in the climate system by restraining the terrestrial uptake of atmospheric CO2. Surface temperature and precipitation changes alter the direction and magnitude of the nitrogen cycle in response to global climate change (Brzostek et al., 2012). Concerns have been expressed regarding "the capacity of natural ecosystems to continue to serve as sinks for atmospheric CO2 over decades to come" (Crous et al., 2010).

About 10 billion anthropogenic tons of carbon (C) are released globally into the environment every year, which contributes to the reduction of species and ocean acidification (Harris, 2012). Oceans have absorbed about one third of global anthropogenic CO2 concentrations which has contributed to decreased surface water pH (i.e., acidification) (Paeth et al., 2009). Global average surface pH has declined by 0.1 since the industrial revolution in the 1800s and a 0.1 pH decrease has been recorded in the upper ocean. The average upper ocean pH is projected to decrease by an average of 0.3 by 2100 (Williamson & Turley, 2012) provided no reduction of global CO2 emission takes place (Paeth et al., 2009). "Such ocean acidification may have ecological and biogeochemical impacts that last for many thousands of years." The degree of future ocean acidification will be largely determined by atmospheric CO2 levels. "Serious consequences for ecosystems...seem inevitable on decadal-to-millennial time scales if CO2 emissions continue on current trajectories." A burning of all planetary fossil fuels will contribute to a 0.7 pH surface water decline compared to pre-industrial times (Williamson & Turley, 2012).

11.6 Climate Change & Human Health

Climate change has been recognized as "the biggest global health threat of the 21st century" (Thomas et al., 2012), posing risks to present and future generations (Goes et al., 2011).

Global climate change is projected to adversely affect human health. Health-related consequences from global climate change include increased illness and mortality resulting from an accelerating rate of heat waves, infectious disease, physical impairment, malnutrition as a result of drought, increased allergic rhinitis due to increased pollen levels, growth in indoor pests and allergens as a consequence of elevated indoor dampness levels, lower indoor air quality due to reduced natural ventilation and psychiatric disorders as a consequence of flooding (Peel et al., 2013). The most vulnerable to climate change populations are those who have already been impacted by climate-change induced disease. Climate change poses risks to human health mainly by intensifying existing health issues (Neira, 2014).

11.6.1 Mental Health

Climate change affects the economic, environmental and social conditions that influence mental health. Greater weather variability and extreme weather events are projected to add to mental health risks, including post traumatic stress disorder (PTSD), anxiety, dementia, mood affective disorders and suicide (Thomas et al., 2012).

11.6.2 Hunger & Malnutrition

Climate change has an impact on food, water and nutrition security (Neira, 2014). Climate change and more particularly solar radiation, temperature and precipitation variations "have the potential to significantly disrupt food production" and thus increase the risk for hunger and malnutrition, especially in the developing nations which are particularly prone to the adverse health effects resulting from climate change (Thomas et al., 2012).

11.6.3 Foodborne & Waterborne Infectious Disease

Climate change may contribute to the development of foodborne and waterborne infectious disease (Schijven et al., 2013). Bacteria, viruses, parasites and protozoa are transmitted via food. 6 to 80 million people are affected by foodborne disease each year in the U.S., contributing to about 9,000 deaths (Park et al., 2006). Increased temperatures and increased frequency of droughts may lead to the rise of certain pathogens and infectious diseases while elevated annual precipitation patterns contribute to an overall increase in infection risk from all pathogen agents. Studies have found an association between temperature, food poisoning, ear and wound infections. Findings from studies also establish a correlation between increased precipitation and outbreaks related to drinking water. Depending on the pattern, climate change affects foodborne and waterborne pathogenic pathways differently (Schijven et al., 2013).

Higher temperatures are expected to result in greater dehydration episodes, increased risk of renal, allergic, diarrhoeal and infectious disease, heat cramps, syncope, stroke, exhaustion and death (Thomas et al., 2012). Each year, more than one million people in the U.S. suffer from bacterial diarrhea (Price et al., 2007). Approximately 1.5 million children under 5 years of age die from diarrhoea each year. Maternal exposure to higher temperatures has been associated with increased risk of preterm birth and stillbirth (Thomas et al., 2012).

11.6.4 Immune System

Solar ultraviolet radiation (UVR) consists of UVC, UVB and UVA radiation. UVR is the most significant germicide in the environment, inactivating disease-causing micro-organisms in sea and fresh water systems. Climate-induced changes in solar UVR could further impact the immune system. Suppression of the immune system enhances vulnerability to cancer and infection while over-activity could contribute to allergy and autoimmune disease (Thomas et al., 2012). Vulnerability is generally defined as "the weakness of the exposed system" (Silva et al., 2014). DNA damage, cell death and inflammation represent acute responses to elevated doses of UVR exposure.

Skin tumors induced by UVR exposure are on an upward trend in a number of countries in spite of "strong...sun protection programmes" (Thomas et al., 2012).

11.7 Climate Change & Ecosystems

It is predicted that climatic and ecological conditions "outside historical norms" would affect large areas of the plant's surface (Paeth et al., 2009). Climate change is projected to impact biodiversity and ecosystems, including aquatic ecosystems, in the decades to come (Bull et al., 2013). Climate change is projected to be the leading cause for loss of biodiversity in the 21st century (Paeth et al., 2009).

11.7.1 Migration of Species

Climate has an effect on "all aspects of the life cycle of a species" (Ouvrier et al., 2012). Climate change contributes to habitat loss and extinction of species unable to migrate or adapt (Paeth et al., 2009). Climate change has historically contributed to the migration of species to areas with more suitable climates (Nunez et al., 2013).

Climate change has contributed to the global migration of plant species. It is uncertain as to whether plant species will be able to "keep pace with future climate change." Climate change in combination with high rates of deforestation may eliminate climate analogs and/or increase the distance between analogs. Tropical lowland rainforests are particularly vulnerable to high climate change velocities and rates of deforestation. Amazon climate change models project that the distance between current Amazon rainforests and their closest (2050) climate analogue is almost 300 km. If precipitation is included in the model, the distance increases by more than 50% to over 475 km. If deforested areas which act as barriers to species movements are incorporated in the model, 30% to 50% of Amazon rainforests will have no climate analog in tropical South America. Other estimates predict that by 2100 distances to climate analogs will exceed 1000 km for the majority of lowland Amazon rainforests which is equivalent to migration rate of over 10 km per year. Such projections are likely to be "overly conservative" (Feeley, et al., 2012).

It is imperative that areas with different temperatures are connected to allow for the migration of species and hence the conservation of species diversity. Anthropogenic obstacles such as land areas occupied by human beings have the potential to obstruct the migration path of species (Nunez et al., 2013). Climate change affects biodiversity and species range, yet its integration into conservation management strategies is challenging and is therefore rarely taken into consideration (Maalouf et al., 2012).

11.7.2 Extreme Events

The increase of the frequency and intensity of extreme events (i.e., droughts) are climate change manifestations with possible significant impacts on ecosystems (Maalouf et al., 2012). Anthropogenic forest land alterations in addition to extreme weather events could impact ecological interactions and contribute to the reduced ability of carnivores to disperse plant and tree seeds due to a shift from plant-based towards birds and small mammals-based diet (Zhou et al., 2013).

11.7.3 Drylands, Flora & Fauna

Climate models predict increased temperatures and altered precipitation treatment in the next 30-100 years for the majority of drylands in the world. Rising temperatures and more frequent summer precipitation rates in drylands as a result of global and regional climate change could lead to loss or decline of biocrust (i.e., decline in DNA concentration), impaired soil fertility, increased rates of erosion and dust production (Johnson et al., 2012). Further, climate change threatens the continual existence of flora and fauna in the U.S. and worldwide (Povilitis & Suckling, 2010).

11.7.4 Fish & Zooplankton

Climate-induced river streamflow and sediment supply variations could contribute to stream surface alterations and thus impact threatened and endangered fish species (Neupane & Yager, 2013). The Northern California Current Ecosystem (NCCE) is exposed to seasonal climate shifts. Climate change affects zooplankton biomass, "key prey resources in the NCCE pelagic food web" (Francis et al., 2012).

11.7.5 Sea Level Rise (SLR)

Sea level rise (SLR) is perceived as a "growing, certain and prominent consequence of anthropogenic climate change." The Intergovernmental Panel on Climate Change (IPCC) estimated that the mean global sea surface has increased by 1.8 mm per year between 1961 and 1993 and 3.1 mm per year from 1993 until 2003. More current estimates demonstrate that SLR reached 3.4 mm per year between 1995 and 2010. The average SLR rate during the past 3000 years was 0.1-0.2 mm per year. The IPCC estimates a global SLR of 0.18-0.59 m between 1990 and 2100 (Scott et al., 2012). Current estimates range between 0.5 m to several meters prior to the end of the twenty-first century (Wetzel et al., 2012). The majority of estimates project 0.5-1.9 m SLR during the 21st century although greater rates are also anticipated under certain circumstances such as decreased solar albedo resulting from the melting of ice sheets. Growing evidence reveals that the Antarctic and Greenland ice sheets have been increasingly losing mass during the past two decades (Scott et al., 2012). Current estimates suggest that SLR is taking place at a 60% faster rate than the projections established by the IPCC.

SLR is projected to result in primary effects (i.e., land

loss from erosion and inundation) as well as secondary effects (i.e., increased rates of mortality from infectious disease following wars; ecological impacts as a result of human displacement). "Human displacement from SLR has enormous economic, political, and medical implications" (Wetzel et al., 2012).

11.7.6 Landfast Sea Ice

Landfast sea ice is defined as still sea ice, attached to ice walls or fronts, to the shore, between icebergs or shoals. Landfast sea ice occurs seasonally in the Arctic, ranging 5-50 km off the Alaska coast to several hundred kilometers off the Siberia coast. The likelihood for solar heating increases as a result of open water regions at the interface between mobile and landfast sea ice. Landfast sea ice is an essential biological habitat for polar bears, seals as well as for the lower trophic levels of the ecosystem. Estimates suggest that the duration of the landfast season is becoming shorter in a number of areas of the Canadian Arctic. Open water areas further contribute to the increase of ice edge vulnerability to wind waves and swell. It is a well-established fact that mobile sea ice in the Northern Hemisphere has undergone "significant reductions in both its thickness and its extent" (Galley et al., 2012).

11.7.7 Summer Sea Ice

The total Arctic volume of summer sea ice has been decreasing since 1987. In 2007, Arctic summer sea ice retreated "dramatically...shattering the previous record low ice extent set in 2005 by 23%." The increasing volume of summer sea ice loss is the result of global warming in general and surface albedo alterations caused by premature onset of sea ice melt, greater amounts of thin ice, and more open water in particular (Lindsay et al., 2009).

11.7.8 Coastal Areas

A large number of coastal areas will be lost as a result of SLR resulting from global warming (Wetzel et al., 2012). 70% of the world's monitored beaches are receding or eroding as a result of SLR. Amongst many, SLR is projected to affect coastal areas by inundating coastal lands, eroding of beach areas, flooding, higher water tables and salinity intrusion into freshwater systems. Studies estimate tens of billions of dollars of losses related to tourism in Florida by 2050 as a result of 0.68 m SLP. More specifically, half of existing beaches, 1362 hotels and motels, 74 airports and 19,000 historic sites would be impacted in Florida. On a similar note, at 1.4 m SLR, damage greater than $2 billion would be incurred at five primary beach areas in California by 2100 (Scott et al., 2012).

11.7.9 Island Species, Land & Terrestrial Biodiversity

The effects of SLR on island species distributions and terrestrial biodiversity were investigated. The study focused on over 1,200 islands in the Pacific and Southeast Asian regions. According to estimates, 3% of coastal land (i.e., habitat and agricultural area) is projected to be inundated as a result of 1 m SLR, 13% under a 3 m SLR, and 32% under a 6 m SLR. A number of species would be at risk of becoming extinct if they are unable to adjust to habitat loss.

Large island land area losses are projected to take place as a result of SLR. The likely effects of SLR on island land loss and its impact on biodiversity under a 1 m, 3 m and 6 m SLR scenario were examined. The study focused on over 12,900 islands (42% of global island area) of the Southeast Asian and Pacific (SEAP). Even under a 1 m SLR, the islands are projected to experience a large area loss as a result of land erosion and water inundation. The total area loss is estimated to amount to the size of Haiti under 1 m SLR scenario, Iceland under 3 m SLR scenario, and New Zealand under a 6 m SLR scenario.

15% to 62% of total island area is likely to be lost under the 1 m to 6 m scenario. Such land loss is predicted to contribute to "reduced habitat connectivity at a significant rate and reduced gene flow between populations." 113 to 67 species inhabiting the most vulnerable island groups are expected to become extinct under a 1 m and 6 m SLR scenario respectively. The findings from the present study are most likely conservative due to the fact that additional aspects of climate change may also have an impact on measured variables (Wetzel et al., 2012).

11.8 Climate Change & Fire Events

Spontaneous and prescribed wildfires contribute to global climate change (Peel et al., 2013). Recent research established a strong correlation between climate conditions and fire in western United States over a period of 34 years (Enache & Cumming, 2009). Climate change is projected to contribute to higher summer temperatures in Western United States which could increase the intensity and frequency of fires in Western United States and hillslope erosion (Neupane & Yager, 2013). Climate affects erosion processes on Earth's surface (Herman et al., 2013).

11.9 Climate Change Solution

The international sector has not succeeded to adequately address the danger of global climate change for the past two decades (Preston, 2013). The real solution to climate change is embedded in the root cause of climate change (Victor, 2008). The risks of climate change can be lessened by implementing programs aimed at reducing poverty, preventing health problems, decreasing GHG emissions and atmospheric pollution (Neira, 2014).

"Environmental climate models live in sublime paralyzes,

yielding to the hypothesis and its severity."

"We journey into the labyrinth, but we lose our sense of direction."

By Sosé Gjelaj

Discussion

In line with Sosé's theory, it could be concluded that the U.S. refuses to become a signatory of the Kyoto Accord as fossil fuel combustion is a profitable niche. The created by fossil fuel combustion air pollution translates into a chain reaction of profit (i.e., industrial; chronic disease, etc.).

On a similar note, the South African government disregards its commitment to reduce GHG emission by 40% by the year 2050 as soon as public demand for energy increases. It readily permits the additional creation of two power plants to respond to such rising demand. The only purpose such commitment serves is therefore to deceive the public that the government is concerned about air pollution when in fact all that it is concerned about is profit.

Most interestingly, it appears as if the public is being held accountable for requiring greater energy resources and therefore for the rising GHG emission levels. As will become apparent in the "Downward Acceleration" chapter, it is because of the unequal distribution of wealth on the first place that humanity must place a tremendous effort to grow the centralized economy if it is to even survive. Such effort logically translates into greater energy expenditure. The "Oil Solicit Industry" reveals the same pattern of transfer of responsibility. Pinpointing who to blame prevents the focus on revealing the true face of crime. Thus, profit continues to accrue for governments and industries.

The commitment to reduce GHG emission is further bound to failure as it is only targeting the symptoms of the problem rather eradicating the problem from its root. It is a quick-fix solution and a part of the deceit. If fossil fuel combustion is the source of the problem, the solution to the problem, as also confirmed in "From Energy Crisis to Free Energy," is the utilization of environmentally friendly energy sources, not the treatment of the symptoms of the problems. Yet, tackling symptoms is sufficient to deceive the public that governments are concerned about air pollution and are taking measures to improve air quality. An effective deceit translates into long-term profit as humanity is assured that governments are in control of the particular situation when in fact they are not.

Lastly, accelerating climate change rates are depriving plant species and trees from successfully migrating to climate analogues while rising CO2 levels are contributing to acidic water systems and endangering aquatic life. One can only imagine the extent of profit and power that is to be generated from the creation of such environmental problems.

References

Bader, M. K.-F., Leuzinger, S., Kell, S. G., Siegwolf, R. T. W., Hagedorn, F., Schleppi, P. & Korner, C. (2013). Central European hardwood trees in a high-CO2 future: Synthesis of an 8-year forest canopy CO2 enrichment project. *Journal of Ecology, 101,* 1509-1519.

Barron-Gafford, G. A., Scott, R. L., Jenerette, G. D., Hamerlynck, E. P. & Huxman, T. E. (2013). Landscape and environmental controls over leaf and ecosystem carbon dioxide fluxes under woody plant expansion. *Journal of Ecology, 101,* 1471-1483.

Brzostek, E. R., Blair, J. M., Dukes, J. S., Frey, S. D., Hobbie, S. E., Melillo, J. M., Mitchell, R. J., Pendall, E., Reich, P. B., Shaver, G. R., Stefanski, A., Tjoekler, M. G. & Finzi, A. C. (2012). The effect of experimental warming and precipitation change on proteolytic enzyme activity: Positive feedbacks to nitrogen availability are not universal. *Global Change Biology, 18,* 2617-2625.

Bull, J. C., Mason, S., Wood, C. & Price, A. R. G. (2013). Benthic marine biodiversity patterns across the United Kingdom and Ireland determined from recreational diver observations: A baseline for possible species range shifts induced by climate change. *Aquatic Ecosystem Health & Management, 16*(1), 20-30.

Crous, K. Y., Reich, P. B., Hunter, M. D. & Ellsworth, D. S. (2010). Maintenance of leaf N controls the photosynthetic CO2 response of grassland species exposed to 9 years of free-air CO2 enrichment. *Global Change Biology, 16,* 2076-2088.

Dhansay, T., de Wit, M. & Patt, A. (2014). An evaluation for harnessing low-enthalpy geothermal energy in the Limpopo Province, South Africa. *South African Journal of Science, 110*(3/4), 1-10.

Dilling, L. & Hauser, R. (2013). Governing geoengineering research: Why, when and how? *Climatic Change, 121,* 553-565. Enache, M. D. & Cumming, B. F. (2009). Extreme fires under warmer and drier conditions inferred from sedimentary charcoal morphotypes from Opatcho Lake, central British Columbia, Canada. *The Holocene, 19*(6), 835-846.

Farahani, J. V. (2014). Man-made major hazards like earthquake or explosion; case study Turkish mine explosion. *Iranian Journal of Public Health, 43*(10), 1444-1450.

Feeley, K. J., Malh, Y., Zelazowki, P. & Silman, M. R. (2012). The relative importance of deforestation, precipitation change, and temperature sensitivity in determining the future distributions and diversity of Amazonian plant species. *Global Change Biology, 18,* 2636-2647.

Ferraro, A. J., Highwood, E. J. & Charlton-Perez, A. J. (2014). Weakened tropical circulation and reduced precipitation in response to geoengineering. *Environmental Research*

Letters, 9, 1-7.

Francis, T. B., Scheuerell, M. D., Brodeur, R. D., Levin, P., Ruzicka, J. J., Tolimieri, N. & Peterson, W. T. (2012). Climate shifts the interaction web of a marine plankton community. *Global Change Biology, 18*, 2498-2508.

Galley, R. J., Else, B. G. T., Howell, S. E. L., Lukovich, J. V. & Barber, D. G. (2012). Landfast sea ice conditions in the Canadian Arctic: 1983 - 2009. *Arctic, 65*(2), 133-144.

Goes, M., Tuana, N. & Keller, K. (2011). The economics (or lack thereof) of aerosol geoengineering. *Climatic Change, 109*, 719-744.

Goldblatt, C. & Watson, A. J. (2012). The runaway greenhouse: Implications for future climate change, geoengineering and planetary atmospheres. *Philosophical Transactions of the Royal Society, 370*, 4197-4216.

Harris, A. (2012). The danger of playing God. *Engineering & Technology, 7*(4), 36-39.

Herman, F., Seward, D., Valla, P.G., Carter, A., Kohn, B., Willett, S. & Ehlers, T. A. (2013). Worldwide acceleration of mountain erosion under a cooling climate. *Nature, 504*, 423-436.

Hou, Y., Wang, S., Zhou, Y., Yan, F. & Zhu, J. (2013). Analysis of the carbon dioxide concentration in the lowest atmospheric layers and the factors affecting China based on satellite observations. *International Journal of Remote Sensing, 34*(6), 1981-1994.

Johnson, S. L., Kuske, C. R., Carney, T. D., Housman, D. C., Gallegos-Graves, La V. & Belnap, J. (2012). Increased temperature and altered summer precipitation have differential effects on biological soil crusts in a dryland ecosystem. *Global Change Biology, 18*, 2583-2593.

Lindsay, R. W., Hang, J. Z., Chweiger, A. S., Steele, M. & Stern, H. (2009). Arctic sea ice retreat in 2007 follows thinning trend, *Journal of Climate, 22*, 165-176.

Maalouf, J.-P., Le Bagousse-Pinguet, Y., Marchand, L., Bachelier, E., Touzard, B. & Michalet, R. (2012). Integrating climate change into calcareous grassland management. *Journal of*

Applied Ecology, 49, 795-802.

Neira, M. (2014). The 2014 WHO conference on health and climate. *Bulletin of the World Health Organization, 92,* 546.

Neupane, S. & Yager, E. M. (2013). Numerical simulation of the impact of sediment supply and streamflow variations on channel grain sizes and Chinook salmon habitat in mountain drainage networks. *Earth Surface Processes and Landforms, 38,* 1822-1837.

Nunez, T. A., Lawler, J. J., McRae, B. H., Pierce, D. J., Krosby, M. B., Kavanagh, D. M., Singleton, P. H. & Tewksbury, J. J. (2013). Connectivity planning to address climate change. *Conservation Biology, 27*(2), 407-416.

Ouvrier, S. J., Holland, M., Stroever, J., Barbraud, C., Weimerskirch, H., Serreze, M. & Caswell, H. (2012). Effects of climate change on an emperor penguin population: Analysis of coupled demographic and climate models. *Global Change Biology, 18,* 2756-2770.

Ozymy, J. & Jarrell, M. L. (2011). Upset over air pollution: Analyzing upset event emissions at petroleum refineries. *Review of Policy Research, 28*(4), 365-381.

Paeth, H., Born, K., Girmes, R., Podzun, R. & Jacob, D. (2009). Regional climate change in tropical and Northern Africa due to greenhouse forcing and land use changes. *Journal of Climate, 22,* 114-132.

Park, D-K., Bitton, G. & Melker, R. (2006). Microbial inactivation by microwave radiation in the home environment. *Journal of Environmental Health, 69*(5), 17-24.

Peel, J. L., Haeuber, R., Garcia, V., Russell, A. G. & Neas, L. (2013). Impact of nitrogen and climate change interactions on ambient air pollution and human health. *Biogeochemistry, 114,* 121-134.

Pendergrass, A. G. & Hartmann, D. L. (2014). Changes in the distribution of rain frequency and intensity in response to global warming. *Journal of Climate, 27,* 8372-8383.

Povilitis, A. & Suckling, K. (2010). Addressing climate change threats to endangered species in U.S. recovery plans. *Conservation Biology, 24*(2), 372-376.

Preston, C. J. (2013). Ethics and geoengineering: Reviewing the moral issues raised by solar radiation management and carbon dioxide removal. *Wiley Interdisciplinary Reviews: Climate Change, 4,* 23-37.

Price, L. B., Lackey, L. G., Vailes, R. & Silbergeld, E. (2007). The persistence of fluoroquinolone-resistant *Campylobacter* in poultry production. *Environmental Health Perspectives, 115*(7), 1035-1039.

Robock, A., Marquardt, A., Kravitz, B. & Stenchikov, G. (2009). Benefits, risks, and costs of stratospheric geoengineering. *Geophysical Research Letters, 36,* 1-9.

Sang, Y.-F. (2013). Wavelet entropy-based investigation into the daily precipitation variability in the Yangtze River Delta, China, with rapid urbanizations. *Theoretical & Applied Climatology, 111,* 361-370.

Schijven, J., Bouwknegt, M., de Roda Husman, A. M., Rutjes, S., Sudre, B., Suk, J. E. & Semenza, J. C. (2013). A decision support tool to compare waterborne and foodborne infection and/or illness risks associated with climate change. *Risk Analysis, 33*(12), 2154-2167.

Scott, D., Simpson, M. C. & Sim, R. (2012). The vulnerability of Caribbean coastal tourism to scenarios of climate change related sea level rise. *Journal of Sustainable Tourism, 20*(6), 883-898.

Silva, R., Martinez, M. L., Hesp, P. A., Catalan, P., Osorio, A. F., Martell, R., Fossati, M., da Silva, G. M., Marino-Tapia, I., Pereira, P., Cienguegos, R., Klein, A. & Govaere, G. (2014). Present and future challenges of coastal erosion in Latin America. *Journal of Coastal Research, 71,* 1-16.

Strachan, P. A., Lala, D. & Malmborg, F. V. (2006). The evolving UK wind energy industry: Critical policy and management aspects of the emerging research agenda. *European Environment, 16,* 1-18.

Thomas, P., Swaminathan, A. & Lucas, R. M. (2012). Climate

change and health with an emphasis on interactions with ultraviolet radiation: A review. *Global Change Biology, 18,* 2392-2405.

van Wezel, A. P., Kruitwagen, S. & Maas, R. (2006). Policy profile: How Dutch environmental policy contributes to meet European environmental standards; Dutch environmental balance. *European Environment, 16,* 45-52.

Victor, D. G. (2008). On the regulation of geoengineering. *Oxford Review of Economic Policy, 24*(2), 322-336.

Wei, W., Chen, L. & Fu, B. (2009). Effects of rainfall change on water erosion processes in terrestrial ecosystems: A review. *Progress in Physical Geography, 33*(3), 307-318.

Wetzel, F., Kissling, W. D., Beissmann, H. & Penn, D. J. (2012). Future climate change driven sea-level rise: Secondary consequences from human displacement for island biodiversity. *Global Change Biology, 18,* 2707-2719.

Williamson, P. & Turley, C. (2012). Ocean acidification in a geoengineering context. *Philosophical Transactions of the Royal Society, 370,* 4317-4342.

Wilson, J., Tyedmers, P. & Spinney, J. E. L. (2013). An exploration of the relationship between socioeconomic and well-being variables and household greenhouse gas emissions. *Journal of Industrial Ecology, 17*(6), 880-891.

Zhou, Y., Newman, C., Chen, J., Xie, Z. & Macdonald, D. W. (2013). Anomalous, extreme weather disrupts obligate seed dispersal mutualism: Snow in a subtropical forest ecosystem. *Global Change Biology, 19,* 2867-2877.

Chapter 12

GEOENGINEERING

Overview

"Geoengineering" elaborates on geoengineering techniques, policies, regulations, hazards, research and application.

***

"Problems are rarely as they first appear" (Edwards & Gibeau, 2013).

12.1 Geoengineering

Geoengineering is defined as "the deliberate large-scale manipulation of the planetary environment to counteract anthropogenic climate change" (Dilling & Hauser, 2013). The main goal of geoengineering is to "moderate surface temperature" (Baughman et al., 2012). Geoengineering is an alternative mitigation strategy to global warming as efforts to reduce anthropogenic GHG emissions have not been successful (Ferraro et al., 2011). Geoengineering is often perceived as an intermittent to mitigation method (Preston, 2013).

The climate system is complex (Dilling & Hauser, 2013). Both geoengineering and the Kyoto protocol presume that climate change dynamics are uncomplicated when in reality they are complex. Both geoengineering and the Kyoto protocol presume that "climate change is a 'problem' to be 'fixed' rather than a condition to be managed." Such an assumption contributes to an accentuation on "silver bullet" simple solutions contrasted to management strategies embedding social, economic and technological responses. As presently conceived, geoengineering is a solution to a specific challenge thus separating climate change from other domains. In reality, however, geoengineering is projected to fundamentally affect a number of additional systems (i.e., ecosystems; biodiversity; agriculture; natural elemental cycles; social; economic; technological; demographic, etc.). The changes on such systems are expected to be significant and possibly highly damaging. Altering monsoon rain patterns in Asia, for instance, could contribute to famine and mass migration, which could result in substantial social and political instability (Allenby, 2012).

12.1.1 Approaches

There are two main branches of geoengineering: solar radiation management (SRM) and carbon dioxide removal (CDR) (Preston, 2013).

12.1.1.1 Solar Radiation Management (SRM)

SRM methods are designed to reduce the amount of solar irradiance reaching earth via the increase of the albedo or reflectivity of the lower, mid and upper atmosphere or of the ocean or land surfaces (Williamson & Turley, 2012). In other words, radiation management is the process of preventing sunlight from reaching Earth (Morton, 2007). SRM is projected to affect precipitation and the hydraulic cycle among other climatic variables (Dilling & Hauser, 2013). SRM-induced cooling of the planet would contribute to reduced precipitation and evaporation rates. Some regions will be rendered more protected than others, "creating local winners and losers." "SRM could alter the global climate within months" (Keith et al., 2010).

A great number of uncertainties relative to solar geoengineering exist and need to be addressed when evaluating the risks and benefits of geoengineering. Its implementation is "unlikely to be so simple" (Kravitz et al., 2014). SRM would not allow for the "return to the exact conditions of...pre-industrial climate." A drastic discontinuation of SRM would contribute to a rapid increase in temperatures resulting from elevated CO2 levels. A "lock-in" of SRM techniques may result, preventing alternative weathering options from being considered. SRM methods concern "only symptoms rather than causes," fail to address ocean acidification and allow CO2 to remain in the atmosphere. Examples of SRM include the injection of sulfate particles in the stratosphere, increasing the albedo of marine surfaces and inserting reflective mirrors in space (Preston, 2013).

12.1.1.1.1 Stratospheric Geoengineering

Injecting sulfate aerosol precursors into the stratosphere is a geoengineering method designed to reduce global warming, cool planet earth, stop the melting of sea ice and glaciers, increase terrestrial C sink and reduce the rate of SLR (Robock et al., 2009).

Stratospheric sulfur injection involves the artificial injection of sulfate aerosol particles into the stratosphere. The method is intended to simulate the reflective cooling effect derived from volcanic eruption activity (Baughman et al., 2012). Inserting sulfate particles in the stratosphere in order to increase the albedo of the planet is of great interest as it not only resembles a process already known to scientists, volcanic eruption, but it also allows for sulfate particles to remain in the stratosphere for 1-3 years prior to needing to replenish them. Removing sulfate aerosols from the stratosphere would be challenging once injected (Victor, 2008). The amount of cooling of the planet will depend on the amount of sulfate aerosol injected in the stratosphere and duration of maintaining the sulfate aerosol cloud in the stratosphere (Robock et al., 2009). Its extensive atmospheric lifetime in addition to the unpredictability of the climate response are considered to be risk factors (Baughman et al., 2012).

In other words, two of the risks associated with aerosol geoengineering include the possibility of "failure to sustain the aerosol forcing" as well as the adverse effects associated with aerosol geoengineering forcing. The lifetime of stratospheric aerosols is a couple of years (Goes et al., 2011). Such method is projected to contribute to a global decrease in precipitation, to regionally unequally distributed effects (Ferraro et al., 2014b) or uneven counteracting of climate change, ocean acidification, resulting from the "still-high" atmospheric CO2 levels, risk of droughts and alteration of ecosystem processes (Victor, 2008). Injecting aerosol particles in the stratosphere compromises the ozone layer which could potentially lead to damages to plants, phytoplankton, skin and eyes. The generating capacity of solar power would be impacted by stratospheric geoengineering with sulfate aerosols. Stratospheric geoengineering with sulfate particles is predicted to cause whitening of the sky (Kravitz et al., 2012). There are other possible side effects which may not be known at present, yet "will appear once analysts start looking for them more aggressively" (Victor, 2008).

It would cost the United States (U.S.) military several billion dollars per year to conduct the procedure utilizing existing equipment such as fighter and tanker airplanes. The cost would be much higher if balloons or artillery are used instead. "The military has already manufactured more planes than would be required for this geoengineering scenario". The cost of each plane ranges between $29.9 million and $88.4 million per plane with annual maintenance of either $3.7 million or $4.6 million depending on the airplane and a total of approximately $25 million in operational cost per plane per year (Robock et al., 2009).

"Aerosol geoengineering...can represent severe risks for future generations." Discontinuing aerosol geoengineering is not only projected to result in drastic rates of global temperature increases or rapid global warming, but also in substantial "economic damage" (Goes et al., 2011).

12.1.1.1.2 Cloud Albedo

Cloud albedo, cloud brightening (Baughman et al., 2012) or marine cloud brightening (MCB) is a SRM method designed to counteract anthropogenic global warming via engineering the planet's climate. As a result of an increase in GHGs, anthropogenic aerosol particles intensify the albedo of clouds or "the fraction of solar radiation reflected back to space," counteracting 20-40% of radiative forcing. MCB is designed to enhance low-level marine stratocumulus clouds albedo via the injection of sea water aerosols in them (Wood & Ackerman, 2013). Cloud brightening is considered to be a financially and technologically advantageous geoengineering method compared to other techniques including stratospheric sulfur injection. However, only a couple of modeling simulations have been developed to test its effects to date. Its effects are largely unknown and potentially detrimental.
  Geoengineering via cloud brightening is projected to have adverse impact on the entire climate system. Such effects include changes in deep water, ocean currents and seasonal ice. It is demonstrated that the cooling effect derived from cloud modification would not be sufficient to counteract the climate effects from increasing greenhouse gas emissions, especially in the long-run. Increased Arctic and global warming are projected to continue increasing despite the application of cloud modification techniques (Baughman et al., 2012).

12.1.1.2 Carbon Dioxide Removal (CDR)

The aim of CDR techniques is to reduce global warming by offsetting atmospheric CO2 emissions (Williamson & Turley, 2012). Ocean fertilization, afforestation and large-scale spread of synthetic algae are examples of CDR (Preston, 2013). A CDR method, direct air capture, designed to remove and store atmospheric CO2, "is already in use at pilot scale" under London Convention/London protocol regulations (Williamson & Turley, 2012).

CDR is perceived as more natural, lengthy and lasting approach to addressing global climate change compared to SRM methods (Preston, 2013). CDR techniques alone are unlikely to contribute to international policy target CO2 reduction levels (Williamson & Turley, 2012). A drastic discontinuation of CDR techniques would allow for the rapid excess of GHG emission levels (Preston, 2013).

12.1.2 Governance

The topic of geoengineering has gained increased popularity in the government and media in the past few years (Dilling & Hauser, 2013). The growing interest in geoengineering is projected to create the need for regulations (Victor, 2008).

Climate modification is of interest to "global powers and those invested in the political stability of particular regions" (Preston, 2013). Geoengineering is of interest due to its low cost and the ability of one or a few countries to geoengineer the global climate without cooperation from other nations. Countries can easily "avoid international commitments" and deploy geoengineering unilaterally (Victor, 2008).

Despite the fact that weather modification for military purposes has been banned by the Environmental Modification Convention, climate engineering technique may still be in use "either for hostile purposes" or as a national attempt to lessen the impact of rising temperatures on the population. As an alternative, a wealthy individual may implement geoengineering for the "greater good." Provided geoengineering is privately owned, its benefits could be skewed towards individuals able to pay rather towards the populations that are in most need of them.

Achieving a global consent to geoengineering is perceived as unrealistic due to the variety of competing objectives (Preston, 2013). Only one nation would be required to fund and deploy geoengineering for the global climate to be affected. Cooperation from other nations would not be necessary. Cooperation is much more likely once geoengineering the planet has begun as withdrawing it would contribute to a rapid global warming 20 times greater than the global warming taking place in present times (Victor, 2008). In other words, if geoengineering is to be initiated, it would have to be continued for decades while a decision to stop geoengineering would result in drastic global warming (Robock et al., 2009). "Once the process of geoengineering begins—whether unilateral or collective—it is likely the world will be unable to stop."

Geoengineering the planet is projected to cost $100 billion as opposed to $1 trillion for mitigation. The more climate change is exacerbated, the more side effects resulting from geoengineering would be tolerated, the more the need for interventions due to the complexity of the climate system and the greater the cost. In practice, thereafter, "the geoengineering cocktails that are likely to be deployed will not be cheap." The most innovative minds and organizations tend not to "put complex technologies" that they create "on the shelf." "The way societies usually get something on the shelf, politically and organizationally, is to use it" (Victor, 2008).

12.1.3 Research

The U.S. human subject institutional review board (IRB) requires that conducting research, such as geoengineering research, does not cause significant harm on human populations (Dilling & Hauser, 2013). International governance does not apply to geoengineering experiments as long as it can be demonstrated that "the experiments do not have lasting impacts on regional or global climate and weather" (Wood & Ackerman, 2013).

Geoengineering research is impossible to distinguish from actual geoengineering implementation "which no-one is suggesting should be done at this time" (Dilling & Hauser, 2013). SRM-induced environmental risks cannot be estimated without field testing specific techniques that may be employed (Keith et al., 2010). The scientific method requires a control condition which is not feasible to establish when studying cloud processes as climatic conditions are beyond scientists' control" (Wood & Ackerman, 2013).

A number of scientists believe that the development of capability to conduct research on SRM will decrease the political will to reduce the emissions of GHGs. Few programs for geoengineering have already been initiated (Keith et al., 2010). U.S. national research programs are already conducting geoengineering research. Geoengineering research undertaken in the private sector "is often proprietary, results are often not fully disclosed to the public, and research is not published in the open peer-reviewed literature" (Dilling & Hauser, 2013).

Informed consent is not required for MCB experiments as they are conducted over marine areas, not in proximity to human populations. The application of the precautionary principle is believed to be ethically appropriate as "it demands that the action be consistent with an understanding of the risks involved." Yet, it is not feasible to delineate the significance or risks of consequences of MCB which may be unknown and unintended. The authors "are not advocates of the implementation of MCB" (Wood & Ackerman, 2013).

12.1.4 "The Cold Reality"

Geongineering is already taking place. "Humans are already in the era of geoengineering" (Allenby, 2012). According to Mark Lynas who has been recognized for his contributions in the field of climate change, "We've...changed the colour of the sky with the aerosol loading down in the atmosphere" (Harris, 2012). Geoengineering is "ridden with flaws and unknown side effects" (Victor, 2008). "The cold reality" is that geoengineering strategies "will not 'solve' the climate 'problem;' rather they will redesign major Earth systems...powerfully and unpredictably" (Allenby, 2012).

"The primordial seas are the music of life's jewels."

"The sun worshipers seeking golden opulence in the sun's steady gaze melts our flesh."

By Sosé Gjelaj

Discussion

Similarly to previous examples of deceit as a means to cover the problem-profit connection as depicted by Sosé, geoengineering is being employed without public's awareness or consent. Expert opinion is not necessary to deduce that it is highly unlikely for a reasonable human being to consent to geoengineering if he or she was aware that geoengineering research is just the same as actual implementation of geoengineering while the hazards of geoengineering are not fully known, yet overwhelmingly present. The scope of profit that geoengineering itself generates and the projected profit from evolving problems related to geoengineering can also be induced. There are no regulations pertaining to geoengineering which provides additional evidence for the fact that geoengineering is an artificially created problem designed to generate profit. Deriving from Sosé's theory of interconnectedness, fossil combustion or problem one leads to geoengineering or problem two and altogether, both lead to profit. Similarly to the fluoride and antibiotic use examples discussed earlier and the implementation of GMOs, which will be reviewed under "GMO Invasion," the superficial solution herein (i.e., geoengineering) is in essence another problem generating additional opportunities for profit.

Further, similar to the water and aquaculture sectors which are owned by limited private companies, climate is now also being privatized by few industries or individuals. Whoever implements geoengineering can easily decide to stop its implementation which would lead to global warming twenty times the rate as it is today. In other words, whoever controls the climate holds perceived power over all Earth. The military sector is also implementing geoengineering which is a different example of governmental power gained through the creation of a problem as is in this particular case, from deliberate climate alterations.

Lastly, the "Paradoxical Formation" chapter indicated that the solar ultraviolet radiation "is the most significant germicide in the environment." Solar radiation management techniques block the sun from reaching Earth which leaves water systems in particular increasingly vulnerable to micro-organism growth state. One can only imagine the profit that pharmaceutical and health care industries would generate following the spread of disease from contaminated with micro-organisms water systems.

References

Allenby, B. (2012). A critique of geoengineering. *IEEE Potentials*, *12*, 22-26.

Baughman, E., Gnanadesikan, A., Degaetano, A. & Adcroft, A. (2012). Investigation of the surface and circulation impacts of cloud-brightening geoengineering. *Journal of Climate*, *25*, 7527-7543.

Dilling, L. & Hauser, R. (2013). Governing geoengineering research: Why, when and how? *Climatic Change*, *121*, 553-565.

Edwards, F. N. & Gibeau, M. L. (2013). Engaging people in meaningful problem solving. *Conservation Biology*, *27*(2),

239-241.

Ferraro, A. J., Highwood, E. J. & Charlton-Perez, A. J. (2011). Stratospheric heating by potential geoengineering aerosols. *Geophysical Research Letters, 38*, 1-6.

Goes, M., Tuana, N. & Keller, K. (2011). The economics (or lack thereof) of aerosolgeoengineering. *Climatic Change, 109*, 719-744.

Harris, A. (2012). The danger of playing God. *Engineering & Technology, 7*(4), 36-39.

Keith, D. W., Parson, E. & Morgan, M. G. (2010). Research on global sun block needed now. *Nature, 463*, 426-427.

Kravitz, B., MacMartin, D. G., Leedal, D. T., Rasch, P. J. & Jarvis, A. J. (2014). Explicit feedback and the management of uncertainty in meeting climate objectives with solar geoengineering. *Environmental Research Letters, 9*, 1-7.

Morton, R. (2008). Historical changes in the Mississippi-Alabama barrier-island chain and the roles of extreme storms, sea level, and human activities. *Journal of Coastal Research, 24*(6), 1587-1600.

Preston, C. J. (2013). Ethics and geoengineering: Reviewing the moral issues raised by solar radiation management and carbon dioxide removal. *Wiley Interdisciplinary Reviews: Climate Change, 4*, 23-37.

Robock, A., Marquardt, A., Kravitz, B. & Stenchikov, G. (2009). Benefits, risks, and costs of stratospheric geoengineering. *Geophysical Research Letters, 36*, 1-9.

Victor, D. G. (2008). On the regulation of geoengineering. *Oxford Review of Economic Policy, 24*(2), 322-336.

Williamson, P. & Turley, C. (2012). Ocean acidification in a geoengineering context. *Philosophical Transactions of the Royal Society, 370*, 4317-4342.

Wood, R. & Ackerman, T. P. (2103). Defining success and limits of field experiments to test geoengineering by marine cloud brightening. *Climatic Change, 121*, 459-472.

Chapter 13

SALVATION

Overview

"Salvation" discusses the topics of food naturalness, food contamination, herbicides, insecticides, pesticides and fungicides, their regulation, health effects and application. Integrated pest management, the green revolution, ethical consumption, sustainable agriculture (i.e., organic agriculture; urban agriculture), insect pollinators, agricultural import and export and the role of farmers are additional topics explored in "Salvation." The chapter concludes with an elaboration of food waste and its prevalence on a global level.

***

Health is defined as the absence of illness and the presence of physical, psychological and social well-being (Siipi, 2013). Both natural and artificial factors, such as what people consume, determine human health (Bongyu et al., 2009). "We are what we eat" implies that we are responsible for taking care of our bodies and that we are to feel good about the food choices that we make (De Tavernier, 2012).

13.1 Food Naturalness

Food is considered healthy when it contains essential for the consumer nutrients, when it does not contain poisonous or harmful to the consumer substances and when it is consumed in the right amount by the consumer. Therefore, whether a food is healthy or not is relational to the consumer and his or her nutritional requirements. The healthiness of food is further dependent on the variety of available food. In cases where there are healthier food alternatives, current food choices would be considered comparatively unhealthy. The food choices made by humanity have ecological consequences.

Food naturalness, the number one global food label claim in 2008, is of importance to a large number of consumers. Food naturalness is, however, an illusion, created by advertising companies and food industries. The definition of naturalness is "highly ambiguous." It could be interpreted to imply little human influence, the opposite of artificial, authentic, familiar, normal.

Naturalness as seen as minimal human influence, is controversial. Most of the food that we consume has been processed in some ways by humans and technology (Siipi, 2013). The majority of food present on the market has undergone a number of industrial processes such as fermentation, heat-treatment, smoking, acid-hydrolysis and drying prior to being consumed (Chaudhry et al., 2008). Degree and kind of food processing does not by itself denote level of naturalness. For example, applying methods to reduce meat pathogens contributes to healthier meat consumption while using trans-fats contributes to unhealthy food. Food manufactured with the right amount and kind of processing rather with minimal processing is what determines food naturalness. Authentic foods do not necessarily imply healthier foods compared to their artificial counterparts. For example, the artificial food additive, almond flavor, is considered as healthier or less harmful than its authentic counterpart. Since food is considered natural and authentic when it is pure in a sense that it does not contain any substances not belonging to it, genetically modified (GM) foods are perceived as being unnatural and not authentic. Consumers perceive organic food as the most natural and GM food as the most unnatural.

The majority of advertisements concerning naturalness refer to the term in absolute rather relational terms. Given the ambiguity of the definition of naturalness, any claim of naturalness must contain an argument that explains why the particular food product is considered natural (Siipi, 2013).

13.2 Food Contamination

13.2.1 Mycotoxins & Cereal

Mycotoxins are fungal formulations with toxic characteristics. Some mycotoxins have carcinogenic properties. Mycotoxins were fist found in breakfast cereal in 1977 (Roscoe et al., 2008). Mycotoxins in cereals either form in the field as a result of infection by fusarium or after the harvesting process as a result of inadequate drying or storage conditions (Scudamore et al., 2008). The maximum mycotoxins level established by the European Community in cereal products is 3 ng g^{-1}. The highest recorded level of mycotoxin in breakfast cereal is 108 ng g^{-1}.

Mycotoxin levels in 156 breakfast cereal samples (i.e., rice; wheat; oat; corn) processed in North America and collected from the Canadian market were analyzed. 3% of the samples or 5 cereal products exceeded the European Community mycotoxin maximum limit. Of all cereal kinds, rice-based breakfast cereals were least contaminated with mycotoxins. "The risk assessment should consider the possible effect(s) of exposure to multiple mycotoxins" (Roscoe et al., 2008).

13.2.2 Trichothecenes

T-2 toxin (T-2), HT-2 toxin (HT-2), deoxynivalenol (DON) and nivalenol (NIV) are trichothecenes that develop in the field and are found in cereal grains such as maize, wheat, oats and barley. Trichothecenes cause an array of disorders in animals, including vomiting, weight loss and feed refusal. Trichothecenes were also found to "inhibit protein, DNA and RNA synthesis, and to have immunosupressive and cytotoxic effects." The presence of trichothecenes in infant foods is especially concerning. Nine cereal products were analyzed for the presence of trichothecenes. Eight samples were contaminated with DON with levels exceeding the levels of T-2 and HT-2. Three samples contained four mycotoxins while the remaining of the samples contained a variety of combinations of toxins (Lattanzio et al., 2008).

13.2.3 Ergot Alkaloids & Rye Flour

Fungus produces a mytotoxin called ergot alkaloids (EAs) which cannot be managed with fungicides. EAs contaminate cereal crops and replace cereal grain and seeds with an ergot body known as sclerotium. Cereal products may be contaminated with EA when the sclerotium is harvested with the cereal grains and seeds. Rye crops are most likely to be adulterated with EAs. Ergotism, an EA intoxication characterized by symptoms such as vomiting, abdominal pain, skin irritation, hallucinations, insomnia, convulsions, heart failure, gangrene of extremities, has been a widespread phenomenon throughout history. The most recent outbreak of ergotism occurred in Ethiopia in 2001. Despite the fact that ergots have been successfully removed from large grain supplies through cleaning techniques, ergots continue to be prevalent in grain products. EAs have been found in Danish, German, Canadian and Swiss cereals in levels of up to 4000 µg kg^{-1}. However, the EU has not established any regulatory consumption limits pertaining to grain, including ground cereal such as rye (Storm et al.,2008).

13.2.4 Fungal Pathogens & Beans

Phaseolus vulgaris is an inexpensive common bean plant, containing high levels of protein, minerals, antioxidants, polyphenols, starch, dietary fiber and vitamins. Fungal pathogens may cause between 80% and 100% loss of bean yield. Chemical treatment of fungal disease on bean crops is not always effective. Planting date and the removal of

171

discolored seeds could reduce fungal infection on bean crops (Dube et al., 2014).

13.2.5 PAH & Oil

PAHs are widely spread carcinogenic contaminants, resulting from natural combustion (i.e., volcanoes; forest fires) and anthropogenic activities (i.e., industrial production; engine exhaust; tobacco smoke) (Rodriguez-Acuna et al., 2008). PAHs are found in a number of food products including vegetable oils (Hernandez-Poveda et al., 2008) such as virgin olive oil (VOO). VOO can become contaminated with PAHs via atmospheric fallout. Vegetables growing close to highways and industrial zones have 10-fold the levels of PAHs compared to crops rearing in rural areas. The contamination of olives during the growing period is transferred onto the final product. A different route of VOO contamination with PAHs occurs during the extraction process as a result of processing combustion fumes. PAHs levels in vegetable oils could be reduced by improved harvesting and processing methods (Rodriguez-Acuna et al., 2008).

13.2.6 Fire Blight, Apples & Pears

Fire blight, caused by the bacterium Erwinia amylovora, is the most detrimental to apple and pear trees bacterial disease. Fire blight affects apple and pear trees in more than forty countries and contributes to great economic losses within the pear and apple industries. Plant pathologists disagree as to whether the pathogen is spread via the fruit. No studies to date demonstrate such association. However, several countries have formulated regulations prohibiting the import of pears and apples from countries where fire blight is present in endemic proportions (Ordax et al., 2009).

13.2.7 Packaged Salad Greens

Packaged salad greens production is also characterized by technological innovation, low cost and labor, high productivity and profit. Packaged salads are advertised as a healthy and time-saving dietary choice for a busy lifestyle. Dole and Fresh Express control 72% of the packaged green salads market. Sales of packaged salads in the U.S. increased by 560% from 1993 until 1999 and from $2.4 billion to $3.9 billion between 2005 and 2007. Processed greens are consumed by 75% of American households. A spring mix packaged salad costs the consumer up to $11.17 per pound while some California farmers attest that bagged greens companies pay them 25 cents per pound. Companies purchase greens from a variety of sources, mixing them at a processing facility. Chlorine is usually used to kill bacteria in greens, yet the permissible legal amounts of chlorine is insufficient to kill bacteria. Packaged greens have resulted in large-scale outbreaks of foodborne disease. The 2006 E. coli outbreak associated with spinach was followed by the California Leafy Greens Marketing Agreement (LGMA), requiring packaged greens industries to develop their own food safety rules. LGMA rules were created by large production industries "behind closed doors without input from other stakeholders." Despite the fact that wildlife does not constitute a significant threat to food safety, LGMA designated certain wildlife animals as "animals of significant risk," requiring farmers to apply measures to eradicate such animals from cultivation areas (Stuart & Woroosz, 2013).

13.3 Herbicides, Insecticides, Pesticides & Fungicides

The insecticide *Zyklon B*, used in gas chambers during World War II, was one of the first synthetic pesticides commercialized in Germany in 1920. Pesticide production grew rapidly in the 1930s. The 1950s marked the development of synthetic herbicides, including *Agent Orange*, which was applied to Vietnamese forests by Americans from

173

1971 until 1981 as a way to cause the wilting of trees. Fungicides, insecticides and herbicides gained wide popularity by the twentieth century. 76,100 tons of active substances were sold in 2004 in France, the third largest pesticides consumer in the world while 2.3 billion Euros was spent on pesticides in 2006. Not only have some weed crops, pests and crop diseases resisted the quick-fix, miracle solutions which indicates a degree of "failure," but pesticides have also been found to pose environmental and public health risks. Studies with animals have demonstrated neurotoxic, cancerous and endocrine system impairment properties of pesticides, rendering the need for further investigation of similar impacts on human beings. Studies with human beings have established a link between certain types of cancer and pesticide use, especially in children. Despite the fact that the French Ministry of Agriculture devised a plan in 2010 to reduce pesticide use by half by 2018, the initiative cannot be successful unless alternative crop cultivation methods are set in place first (Leo & Pintureau, 2013).

13.3.1 Regulation

Congress passed the Food Quality Protection Act of 1996 which imposes upon the Environmental Protection Agency (EPA) a duty to test pesticides for their impact on the endocrine system. Congress afforded EPA a period of three years to begin testing. EPA began mandating testing fifteen years following the enactment of the Food Quality Protection Act or ten years after the deadline set by Congress. "Most disturbingly," EPA's testing policy permits pesticide companies to "submit outdated testing data" (der Mude, 2011).

13.3.2 Health Effects

13.3.2.1 General Pesticides

Strong associations between pesticide exposure and depression have been established in previous studies. Study findings demonstrate that both cumulative (chronic) and acute exposure to pesticides may contribute to the development of physician-diagnosed depression among pesticide applicators. Depression was also found to be linked to chronic pesticide exposure in the absence of pesticide-induced poisoning diagnosed by a physician (Beseler et al., 2008).

13.3.2.2 Chlorpyrifos

Chlorpyrifos is the most commonly used insecticide in the U.S. Increased risk of glioma and non-Hodgkin lymphoma amongst pesticide applicators have been linked to chlorpyrifos exposure. The mortality rate amongst 55,071 Iowa and North Carolina pesticide applicators was investigated. The relative risk of mortality amongst applicators exposed to chlorpyrifos was slightly less than the relative risk of mortality amongst pesticide applicators not exposed to chlorpyrifos. However, suicide (possibly resulting from neurobehavioral impairments) as well as non-motor vehicle accidents rates, were greater amongst chorpyrifos applicators. Increased rates of suicide among pesticide applicators have been reported in previous studies, yet not in others. Similarly, previous studies have found a relationship between chlorpyrifos use, Parkinson disease, lung and brain cancer while the present study did not establish such associations (Lee et al., 2007).

13.3.2.3 Carbofuran

Carbofuran is an insecticide used on food crops such as corn, tobacco, alfalfa and rice. The U.S. uses 5 million pounds of carbofuran per year, 48% of which are applied to corn crops. About half of African American women and their newborn children in the U.S. were noted to have discernible levels of carbofuran in their blood plasma and umbilical cord blood. Carbofuran was found to induce mutagenic activity in strains of Salmonella typhimurium, chromozomal aberrations and micronucelus formations in mice. Increased lung cancer was further demonstrated in human beings exposed to carbofuran. Application of pesticides containing carbofuran, but not exposure to carbofuran specifically, was associated with increased risks of non-Hodgkin lymphoma (NHL). There is substantial evidence demonstrating the carcinogenic properties of nitrosated carbofuran (Bonner et al., 2005).

13.3.2.4 Glyphosate

Herbicides are broadly used for forestry, agricultural and residential weed management (Acquavella et al., 2004). Introduced in the 1970s, glyphosate is a widely used herbicide typically found in Monsanto Company's formulation, *Roundup* (De Roos et al., 2005). Glyphosate was the second most widely applied pesticide in the U.S. in 1999. Its application has been rapidly increasing with the planting of glyphosate-resistant agricultural crops. Regulatory and expert bodies have determined that glyphosate does not serve as a reproductive or developmental toxin, and is not a carcinogen, mutagen or teratogen (Acquavella et al., 2004). Despite the fact that the WHO does not categorize glyphosate as neither a carcinogenic nor a mutagen, some studies show chromosomal aberrations and oxidative stress resulting from a treatment of human lymphocytes with glyphosate. Findings from studies further demonstrate an association between exposure to glyphosate and increased risk of non-Hodgkin lymphoma (NHL). Elevated DNA adducts in mice and mutagenic properties in the Slamonella assay were lastly associated with *Roundup*, but not with glyphosate (De Roos et al., 2005). Studies have further demonstrated the harmful effects of glyphosate on fungal communities (Stefani & Hamelin, 2010).

13.3.2.5 Pesticides & Tea

Tea is the most commonly consumed beverage worldwide. Tea infusion precedes tea consumption. "The residue levels of many pesticides in tea and in its infusion have been reported." Fenvalerate is an insecticide. The residual level of fenvalerate in tea and its transfer properties during tea infusion were estimated. For purposes of the study, fenvalerate was applied on tea crops at two dosages during the dry and wet seasons. Findings revealed a 30-40% loss of residue during tea processing and a 10-30% residue transfer from made tea to infusion for both seasons. 50-70% of the fenvalerate residues remained in the spent tea leaves. It was also noted that infusion-based residue levels were below the limit of detection 7 days after fenvalerate application. A 7-day waiting period is therefore recommended from the time of fenvalerate application to the time for tea plucking (Sharma et al., 2008).

13.3.2.6 Pesticides & Honey

In order to produce honey, honey bees collect nectar and pollen from plants, transform it and store it in honeycombs. Contrary to consumers' perception of honey as a "natural and pure and as free from residues as possible" product, beehives are directly treated with pesticides to control for a parasitic mite called *Varroa jacobsoni*. Additional pesticide residues further accumulate at the honeycombs when honey bees travel to surrounding areas treated with pesticides to collect pollen thus transporting the pesticide residues from the surrounding area to the beehive. The U.S. and the European Union's legislation regarding regulating pesticides in honey is "not always clear and consistent." Maximum residue limits (MRL) have been established for only three types of pesticides directly applied to honeycombs and not to pesticide residues brought from surrounding locations. Some countries such as Italy, Germany and the Netherlands have, however pioneered in establishing MRLs for alternative pesticide residues as well (Blasco et al., 2008).

13.3.3 Integrated Pest Management

Mueller discovered the insecticidal properties of diphenyltricloroethane (DDT) in 1932. His discovery initiated the global use of insecticides in agriculture. The continual use and overuse of insecticides led to resistance to insecticides, pest reemergence, the spread of secondary pests and harmful effects on environmental and human health. Such circumstances contributed to the development of two alternative to chemical pest management paradigms – integrated pest management (IPM) and total pest management (TPM). IPM involves applying pest management measures only when and where it is necessary while TPM implies the continual preventative global application of pest management control strategies (Pringle, 2006).

In 1977, an integrated pest management, designed to reduce agro-environmental and economic risks via a thorough understanding of the ecological plant-environment relationship and the integration of conditions-specific pest management varieties and techniques, was publicly introduced. In the heart of IPM is "the aim to integrate, long-term, all the viable control methods, selecting some rather than others according to the local conditions at any given time, and in this way, preventing any single selection pressure from being too long-lasting." In 2002, IPM had been applied to only 0.4% of cultivated surface land. The difference between conventional and IPM methods is the move away from previously used "defence," "extermination," "enemy," "war-like stance in response to an invasion" pest management methods and into an approach where steps are taken to ensure "care," sustaining "the good health of agricultural crop systems" instead. It has been stated that "the war-like relationship that humans have built up in respect of the universe and nature is a projection of their own war-like relationship with each other; and, for the most part, that is the origin of our ecological problems" (Leo & Pintureau, 2013).

13.4 Green Revolution

The green revolution developed during the late 1950s and throughout the 1960s. It was marked by a shift from traditional or rural to industrialized and capitalized agricultural production process. Following the introduction of the green revolution, hunger was reduced 17% from 1960 until 2000. The incomes of small agricultural producers in developing countries were, however, affected as the green revolution gave birth to large agricultural production enterprises. The rising use of herbicide, insecticides and fertilizers contributed to both environmental deterioration and the concentration of agricultural power within large industries "suspected of neglecting environmental and social issues." Increase in production quantity was undertaken at the expense of production quality. Agricultural output has increased drastically since the 1960s, contributing to climate change. The green revolution was further characterized by the development of pesticide resistance, a reduction in biodiversity and the development of disease due to monocropping. A shift has thereafter taken place from a focus on increased productivity and economic efficiency (green revolution) to agricultural sustainability and environmental well-being (Van Haperen et al, 2012).

13.5 Ethical Consumption

The definitions of ethical consumption vary broadly. Generally, ethical consumption is understood as the conscious choosing of one product over another for reasons other than the product's use-value. Such reasons could exemplify a commitment to environmental, political, spiritual, religious and social values or an objection to unjust market and industrial practices. Companies committed to ethical products have reported a general increase in sales. Other businesses attest that focusing on ethics and sustainability, environmental and social prosperity, improves the economic bottom line. Ethical consumption has gained increasing popularity within the food and agricultural sectors with organic and local food movements leading the way (Long & Murray, 2013).

13.6 Sustainable Agriculture

Sustainable agriculture is proposed to replace conventional agriculture. Conventional agriculture is in its core unsustainable and is primarily being supported by major seed industries and not by the public. Organic and local farming which are part of sustainable agriculture, have been publicly supported instead. Sustainable agriculture has been described as animal and plant cultivation and production methods that ensure the satisfaction of human food and fiber needs, economical and environmental prosperity, improvement of the quality of life for farmers and humanity as a whole. Sustainable agriculture has been equated to "civic agriculture" in that "what is at stake in agriculture is, above all, what is at stake for society in general." In 2007, the Food and Agriculture Organization (FAO) proposed organic farming as the only type of sustainable agriculture that can satisfy the demand for food worldwide while causing minimum environmental impact. A call to integrate all types of sustainable agriculture and not only organic farming as part of the FAO initiative has been set forth (Leo & Pintureau, 2013).

13.7 Organic Agriculture

Rudolf Steiner established the spiritual foundations of organic farming within his philosophy called Anthroposophy (Schosler et al., 2013). Organic agriculture was established by farmers in the 1960s (Long & Murray, 2013) and began gaining popularity in the 1970s (Van Haperen et al., 2012). The organic movement began growing exponentially in the 1980s due to the rise of food-related public health and environmental concerns (Long & Murray, 2013). Organic agriculture was cultivated on 31 million hectares of surface area worldwide in 2006 (Leo & Pintureau, 2013). The U.S. has the largest organic market with sales amounting to $26.7 billion in 2010, a 7.7% increase from 2009. "With the rapid growth of the organic sector, large producers and retailers have come to dominate the market." 25% of the 69% of U.S. consumers of organic produce purchase organic items weekly (Long & Murray, 2013).

The organic approach embeds the principles of care for natural resources, fairness, non-appropriation, agro-biodiversity, holistic, stable, resilient and sustainable agriculture (Van Haperen et al., 2012). Resilience is generally defined as "the capacity of a system to experience shocks...while essentially retaining the same function, structure, and feedbacks" (Silva et al., 2014). "Organic agriculture thus carries anti-biotechnology, anti-capitalism, anti-exploitative, and small-scale tendencies with it" (Van Haperen et al., 2012). Organic food ensures greater sustainability when compared to traditional food consumption choices. Organic farming is a "holistic production management" system which focuses on improving agricultural and ecosystem health while abstaining from the use of synthetic materials. It has been demonstrated that organic agriculture "is capable of feeding the world sustainably," particularly when agricultural practices to mitigate climate change are taken into consideration. Several studies demonstrate that the choice for healthy and natural foods is embedded within a spiritual practice such as mindfulness meditation. Organic consumers view themselves as active participants in the food system.

13.7.1 Organic Food Consumption, The Netherlands

Organic food consumption in the Netherlands is increasing, yet is low compared to other countries in Europe. In depth qualitative interviews were conducted with ten organic consumers from the Netherlands in 2010 to explore their viewpoint on food. Approximately half of the participants were self-employed with a background in art. By choosing to consume organic food the participants were "doing what feels natural" while relying on their senses to point them to the organic food to consume. They felt a certain intuitive connectedness or oneness with the environment which evoked care for self, animals and nature. Participants perceived animals as "sentient fellow creatures" that were treated in cruel ways, like a commodity, within conventional farms. Most participants were vegetarians during the time of the study or at some point in their past. Those presently consuming meat, consumed only organic meat and in moderation. Cooking and eating food was accomplished by a state of awareness of the taste, smell, sensation of the food, by tranquility, absence of thought and attention to the food which ultimately contributed to the experience of joy and peace. Participants believed that humanity is not aware of what is being consumed as "food producers mix substances together and thereby obscure people's choices." Purity or the preservation of the essence of the food being consumed was seen as a priority in food choices. By focusing on their intuition and personal moral values, participants were less likely to be influenced by advertising. Organic food consumers have been depicted as having "a concern which goes beyond the material, a desire for a meaningful life, a moral life, one which is in harmony and balance, a desire for mental peace,..contentment and happiness." Feeling connected with nature is described as a value that needs to be enhanced in an urban aspect through "the development of urban agriculture to enable cities to feed themselves from within or from its neighboring communities" (Schosler et al., 2013).

13.7.2 Transition Period

The transition period from traditional to organic agriculture could last for up to 5 years. Once the organic agricultural system is firmly set in place, it can contribute to crop production as high or even higher than conventional agriculture. Net profit from organic agriculture could further be greater compared to traditional cultivation (Jordan, 2002).

13.8 Urban Agriculture

U.S. consumers are increasingly interested in local, organic and sustainable products which gives rise to urban agriculture in the U.S. Urban agriculture is generally defined as "the growing, processing, and distribution of food and other products through intensive plant cultivation and animal husbandry in and around cities." Urban agriculture incorporates a wide array of farming activities such as growing food on building roofs, decks, community gardens, cultivating crops on vacant industrial land areas, raising animals such as chickens in a backyard to produce a sufficient number of eggs for a couple of families. Health (i.e., organic), environmental (i.e., reduced pollution from transportation and waste product) and economic (i.e., inexpensive produce for urban populations) benefits are associated with urban agriculture (Voigt, 2011).

Land use in the U.S. has been regulated by local and state governments since the 1920s. The most widely used method to control land use is zoning (Ramachandran, 2009). Zoning regulations, "often unintentionally prohibit even the most basic farming activities." For example, zoning regulations may disallow the raising of farm animals in metropolitan areas, growing food on specific land areas, or selling food products from urban farms (Voigt, 2011). Local and state governments further have the authority to "regulate land subdivisions, enforce building code requirements, designate historical districts, as well as encourage desired land use patterns through tax incentives." The Fourteen and Fifth Amendments in the U.S. Constitution protect due process relative to land use. The Fourteenth Amendment does not permit any governmental action that denies "any person of...life, liberty or property without due process of law." The Fifth Amendment disallows the taking of private land for public utilization without fair compensation. Substantive due process concerns governmental justification for depriving a person of liberty, life, or property. "Historically, substantive due process land use claims have fared poorly" (Ramachandran, 2009).

Municipalities have begun to re-consider zoning regulations as a means to facilitate urban farming while minimizing safety, health and nuisance concerns. While such efforts are deemed helpful, they "have been largely piecemeal," leaving "would-be urban farmers confused and discouraged" (Voigt, 2011).

13.9 Agro-Biodiversity

Even though there are 7000 plant species, only 30 species provide 95% of plant-derived proteins and energy intake worldwide. Rice, maize and wheat provide over half of the global calorie and energy intake. Food security depends on preserving the genetic diversity of agricultural plants (Schmidt & Wei, 2006).

13.10 Insect Pollinators

All climate change, habitat transformation, decrease of floral resources, inappropriate application of pesticides, the presence of pests and disease have an effect on the decline of insect pollinators, both managed (i.e., honeybee Apis maillifera) and wild (honeybee Apis spp.). Insect pollinators contribute to improved quality and increased yield of 35% of crops worldwide. Honeybees pollinate crops in many parts of the world. Their general decline is apparent worldwide, specifically in North America and Europe. However, except for the U.S., the number of managed honeybees in particular has been rising worldwide. An increasing number of studies demonstrate that a variety of pollinators provide more effective pollination to crops. Wild insect pollinators were further found to contribute to improved pollination services. Despite the rise in managed honeybees worldwide, the increasing demand for agricultural products and therefore pollination services could exceed honeybees supply. Alternative means to pollination services, improving the health of managed honeybees, expanding the number of wild pollinators are all strategies proposed to ensure adequate pollination services for agricultural crops (Melin et al., 2014).

13.11 Agricultural Import & Export

Virtually all nations depend on agricultural food supplies from other countries in the world. For example, 87% of the plant-derived diet in South Africa and 90% in Europe is imported from other parts of the world (Schmidt & Wei, 2006).

Maize is a main agricultural crop used worldwide for starch, alcohol, fuel, food sweeteners, oil and protein. It is the most widely traded commodity in the world. Therefore, any event that affects the import or export of maize poses a food security risk.

The U.S. is the number one exporter of maize worldwide, exporting 527 million tons of maize to 181 countries from 2000-2009. China, Argentina, France and Brazil are furthermore amongst the largest maize exporters in the world. Japan is the largest maize importer worldwide, importing 170 million tons of maize between 2000 and 2009. Japan is followed by Egypt, Taiwan, Mexico, Spain, Republic of Korea.

Countries such as the U.S., the Netherlands, Mexico, Canada and Germany are amongst the top exporters and top importers of maize. Typically, countries export to and import from a negligible number of other countries, implying that an event that affects maize production and therefore export in one part of the world would impact maize import and hence food security in another part of the world. An increase in maize prices would be a logical consequence of an altered maize commodity trade.

Increase in maize prices was noted in history when the U.S. shifted food production from maize to fuel ethanol production derived from maize. The added increase in energy prices and food production costs as well as changes in weather conditions impacting maize production contributed to a great global increase in food prices. Mexico, importing maize primarily from the U.S., suffered the greatest consequences in terms of food security as a direct result of increased maize prices. Riots, especially amongst the poor Mexican population, arose as a direct result. It is therefore recommended that the largest maize crops producers devise a plan to sustain worldwide maize supply in instances of climate change. On the other hand, the largest importers of maize are advised to prepare plans for food supply in instances when maize production is affected in the major countries exporting maize crops. Two proposed strategies are importing maize from various sources and increasing local food production (Wu & Guclu, 2013).

13.12 Farmers

The food system worldwide has been compromised by "decades of misguided policies" that focused more on food export rather than providing food for humanity. In the past three decades, neoliberal economic policies in a large number of nations contributed to reduced support for farmers who produced food for local markets. Such economic policies further contributed to the public sector reducing food reserves and discontinuing stockpiling of food against famine. This translated into small farmers loosing "a key buyer" which led to the production of less food for the domestic market. The greater part of global food reserves has been deregulated and is "now controlled by the private sector." The outside investing sector can now purchase and sell food in the same manner they buy and sell oil or gold.

Currently, family farmers and peasants produce more than 70% of food in the world. Small farmers tend to produce more food per unit area, utilize less agrochemicals and contribute to a greater extent to a reduction in global warming compared to large corporate farming. Providing peasants and family farmers with support such as access to land, domestic and national markets is necessary (Rosset, 2011).

13.13 Food Waste

In light of rising world population, food security and the environmental damages incurred through food production, food-waste diminution is of prime environmental concern. The general consensus is that between 30% and 50% of food produced for human intake is never consumed (Ridoutt et al., 2014). 1.3 billion tons of food or one third of the global food produced for human consumption is being wasted worldwide each year (Quested et al., 2011).

13.13.1 The U.K.

27.5 million tons of food is wasted in the United Kingdom (UK) each year (Kelleher & Robins, 2013). 20% of waste in landfills in the UK is food waste (Nomura et al., 2011). Household food waste is the "largest contributor" to food waste in the U.K., resulting in 3% of U.K. GHG or the equivalent of 20 million tons of CO_2. Household food waste amounts to 8.3 million tons per year which accounts to a retail value of 12 billion pounds. 5.3 million tons of household food waste is avoidable while 1.5 million tons is unavoidable (i.e., egg shells; apple cores). 5% of drinking water in the U.K. is allocated to household-related avoidable food waste (Quested et al., 2011).

13.13.2 The EU

98 million tons of edible food is wasted in the EU each year. It is predicted that food waste will increase by 42% to a 139 million tons of food waste per year in the next eight years (Kelleher & Robins, 2013). Collected food waste generally has acidic properties, with pH ranging between 4.5 and 6 (Yu et al., 2010).

13.13.3 The U.S.

U.S. households spend approximately $2000/year on food items that are never consumed (Kelleher & Robins, 2013). 34 million tons of food waste are generated in the U.S. on a yearly basis. Most of the food waste is disposed at landfills. 300 million barrels of oil and more than a quarter of freshwater consumption in the U.S. are used in the production of food that is wasted. 14% of GHGs in the U.S. result from the production of food (Dominguez & Moreno, 2012). The Natural Resources Defense Council (NRDC) postulates that 10% of the total U.S. energy budget, 50% of U.S. land and 80% of freshwater consumed in the U.S. are allocated to food production. The yearly waste of food in the U.S. is associated with 10.5 trillion gallons of water waste which is enough to provide water to 500 million people per year (Kelleher & Robins, 2013). Food waste is therefore accompanied by a waste of resources (Dominguez & Moreno, 2012).

13.13.4 Solution

"Reducing the amount of food and drink that is wasted is a key element in developing a sustainable food system" (Quested et al., 2011). Food waste recycling could contribute to the establishment of a healthier environment, yet the promotion of proper household waste recycling is often neglected by local governments (Nomura et al., 2011).

Discussion

"Salvation" provides a number of additional examples in support of various aspects of Sosé's theory.

To begin with, food naturalness is yet another form of deceit designed by industries to mask the true origins and content of food, thus increasing consumption of so-called natural foods and generated profit from such consumption. Further, as will be revealed in the "Nano Manipulation" chapter, it is legal to add nano particles to food without the knowledge of the public. The legal system does not distinguish between nano and natural particles as long as nano particles resemble natural particles on a number of identifiers. If the legal system equates nano food particles to actual natural food particles despite the fact that nano food particles have not been established as safe, it becomes apparent how additional profit could be generated from possible foodborne disease associated with the consumption of foods labeled as natural, yet containing possibly unsafe nano parts.

On a similar note, the food that humanity consumes is contaminated with toxic substances (i.e., mycotoxins; trichothecenes). Food contamination or the problem contributes to the generation of profit from foodborne disease or developed chronic condition. For example, the study by Roscoe et al. (2008) found that 3% or 5 out of 156 of sampled cereal products in Canada exceeded the recommended dose for mycotoxins. The carcinogenic properties of mycotoxins are a sure indicator of the development of cancer, thus profit, as a direct result of consuming contaminated with mycotoxins cereal. At the same time, governments establish regulations (i.e., deceit), yet do not appear to monitor actual levels of food contaminants. Otherwise, 3% of sampled cereal would had not be contaminated with mycotoxins.

Likewise, if the recommended values of chlorine are not sufficient to effectively disinfect packaged salad greens, why would packaged salad greens be commercially available (i.e., problem) unless profit is being generated through human disease upon exposure to contaminants contained within salad packages. It could thus be deduced that, similarly to the East Coast Fever in South Africa, the 2006 E. coli outbreak was also purposefully created in order to generate profit from resulting human disease. The outbreak led to the California Leafy Greens Marketing Agreement (LGMA) which required companies producing packaged greens to develop their own food safety rules. Truth is not revealed to humanity when food safety rules are being developed behind closed doors. There would be no need for safety rules to be established behind closed doors if another poison to humanity is not being created to substitute the initial one. Therefore, disease never ceases to develop and profit generated by the pharmaceutical and the health care sectors never stops to flow.

What is also interesting to notice is that the insecticide *Zyklon B* as well as the herbicide *Agent Orange* were used in gas chambers during World War II and on forests to wilt trees. These same toxins were then applied to cultivated crops despite the implied availability of organic farming substitutes. The health repercussions of such actions are self-explanatory and so is the profit being acquired from resulting human disease. Insecticides are also contributing to the extinction of insect pollinators, which is a drastic problem as crops cannot grow without pollinators. Where is profit going to be generated from if this happens? GM insect pollinators? Highly likely following the established problem-profit pattern.

Furthermore, organic farming was immediately taken

over by large industries as soon as it gained popularity. Deriving from previous examples, not only is it questionable whether corporate organic farming is actually organic, but the centralized profit and power seem only to be transferred from one sector to another, yet never lost. Generally speaking, struggling farmers worldwide are an example of how large agricultural corporations have occupied the agricultural sector. The governments close the doors for farmers (i.e., problem) and open them for agro-businesses (i.e., profit and power). Such train of thought also explains why urban agriculture is literally non-existent. Zoning regulations (i.e., deceit) ensure profit for agro-businesses only. Similarly to the water privatization, aquaculture and climate examples, humanity is left with no other choice but to purchase its food from agro-businesses thus contributing to their imminent expansion and unequal distribution of power.

One of the greatest examples of governmental deceit took place when Congress afforded the EPA three years to test pesticides for their impact on the endocrine system. Not only were testing results not submitted fifteen years after the order, but old data was permitted to be submitted and, indeed, submitted instead. One can conceive of the profit that was generated from poisoning humanity with toxins during the fifteen-year period of delay.

To end, food waste is a staggering problem. It is of no surprise that effective governmental interventions to reduce food waste have not been set in place. The more food is wasted (i.e., problem), the greater, for example, the profit and power for food aid organizations as will be discussed in the "'Non-Profit' Sector and 'Food Aid'" chapters.

References

Acquavella, J. F., Alexander, B. H., Mandel, J. S., Gustin, C., Baker, B., Chapman, P. & Bleeke, M. (2004). Glyphosate biomonitoring for farmers and their families: Results from the farm family exposure study. *Environmental Health Perspectives, 112*(3), 321-326.

Beseler, C. L., Stallones, L., Hoppin, J. A., Alavanja, M. C. R., Blair, A., Keefe, T. & Kamel, F. (2008). Depression and pesticide exposures among private pesticide applicators enrolled in the Agricultural Health Study. *Environmental Health Perspectives, 116*(2), 1713-1719.

Blasco, C., Font, G. & Pico, Y. (2008). Solid-phase microextraction-liquid chromatography-mass spectrometry applied to the analysis of insecticides in honey. *Food Additives and Contaminants, 25*(1), 59-69.

Bongyu, M., Billingsley, G., Younis, M. & Nwagwu, E. (2009). Genetically modified foods and public health debate: Designing programs to mitigate risks. *Public Administration & Management, 13*(3), 191-217.

Bonner, M. R., Lee, W. J., Sandler, D. P., Hoppin, J. A., Dosemeci, M. & Alavanja, M. C. R. (2005). Occupational exposure to carbofuran and the incidence of cancer in the agricultural health study. *Environmental Health Perspectives, 113*(3), 285-289.

Chaudhry, Q., Scotter, M., Blackburn, J., Ross, B., Boxall, A., Caste, L., Aitken, R. & Watkins, R. (2008). Applications and implications of nanotechnologies for the food sector. *Food Additives and Contaminants, 25*(3), 241-258.

De Roos, A. J., Blair, A., Rusiecki, J. A., Hoppin, J. A., Svec, M., Dosemeci, M., Sandler, D. P. & Alavanja, M. C. (2005). Cancer incidence among glyphosate-exposed pesticide applicators in the agricultural health study. *Environmental Health Perspectives, 113*(1), 49-54.

De Tavernier, J. (2012). Food citizenship: Is there a duty for responsible consumption? *Journal of Agricultural & Environmental Ethics, 25*, 895-907.

der Mude, A. V. (2011). Endocrine-disrupting chemicals: Testing to protect future generations. *Environmental Affairs, 38*, 509-535.

Dominguez, S. & Moreno, L. (2012). Tightening the belt by

reducing food waste. *BioCycle, 53*(11), 29-31.

Dube, E., Sibiya, J. & Fanadzo, M. (2014). Early planting and hand sorting effectively controls seed-borne fungi in farm-retained bean seed. *South African Journal of Science, 110*(11/12), 75-80.

Hernandez-Poveda, G. F., Moralez-Rubio, A., Pastor-Garcia, A. & de la Guardia, M. (2008). Extraction of polycyclic aromatic hydrocarbons from cookies: A comparative study of ultrasound and microwave-assisted procedures. *Food Additives and Contaminants, 25*(3), 356-363.

Jordan, C. F. (2002). Genetic engineering, the farm crisis, and world hunger. *BioScience, 52*(6), 523-528.

Kelleher, M. & Robins, J. (2013). What is waste food? *BioCycle, 54*(8), 36-39.

Lattanzio, V. M. T., Solfrizzo, M. & Visconti, A. (2008). Determination of trichothecenes in cereals and cereal-based products by liquid chromatography–tandem mass spectrometry. *Food Additives and Contaminants, 25*(3), 320-330.

Lee, W. J., Alavanja, M. C. R., Hoppin, J. A., Rusiecki, J. A., Kamel, F., Blair, A. & Sandler, D. P. (2007). Mortality among pesticide applicators exposed to chlorpyrifos in the agricultural health study. *Environmental Health Perspectives, 115*(4), 528-534.

Leo, C. & Pintureau B. (2013). Crop protection between sciences, ethics and societies: From quick-fix ideal to multiple partial solutions. *Journal of Agricultural & Environmental Ethics, 26*, 207-230.

Long, M. A. & Murray, D. L. (2013). Ethical consumption, values convergence/divergence and community development. *Journal of Agricultural & Environmental Ethics, 26*, 351-375.

Melin, A., Rouget, M., Midgley, J. J. & Donaldson, J. S. (2014). Pollination ecosystem services in South African agricultural systems. *South African Journal of Science, 110*(11/12), 25-33.

Nomura, H., John, P. C. & Cotterill, S. (2011). The use of feedback to enhance environmental outcomes: A randomised controlled trial of a food waste scheme. *Local Environment, 16*(7), 637-653.

Ordax, M., Biosca, E. G., Wimalajeewa, S. C., Lopez, M. M. & Marco-Noales, E. (2009). Survival of Erwinia amylovora in mature apple fruit calyces through the viable but nonculturable (VBNC) state. *Journal of Applied Microbiology, 107,* 106-116.

Pringle, K. L. (2006). The use of economic thresholds in pest management: Apples in South Africa. *South African Journal of Science, 102,* 201-204.

Quested, T. E., Parry, A. D., Eastel, S. & Swannell, R. (2011). Food and drink waste from households in the UK. *Nutrition Bulletin, 36,* 460-467.

Ramachandran, N. (2009). Realizing judicial substantive due process in land use claims: The role of land use statutory schemes. *Ecology Law Quarterly, 36,* 381-405.

Ridoutt, B. G., Baird, D. L., Bastiaans, K., Darnell, R., Hendrie, G. A., Riley, M. Sanguansri, P., Syrette, J., Noakes, M. & Keating, B. A. (2014). Short communication: A food-systems approach to assessing dairy product waste. *Journal of Dairy Science, 97,* 6107-6110.

Rodriguez-Acuna, R., Del Carmen Perez-Camino, M. D. C., Cert, A. & Moreda, W. (2008). Sources of contamination by polycyclic aromatic hydrocarbons in Spanish virgin olive oils. *Food Additives and Contaminants, 25*(1), 115-122.

Roscoe, V., Lombaert, G. A., Huzel, V., Neumann, G., Melietio, J., Kitchen, D., Kotello, S., Krakalovich, T., Trelka, R. & Scott, P. M. (2008). Mycotoxins in breakfast cereals from the Canadian retail market: A 3-year survey. *Food Additives and Contaminants, 25*(3), 347-355.

Rosset, P. (2011). Preventing hunger: Change economic policy. *Nature, 479,* 472-473.

Schmidt, M. R. & Wei, W. (2006). Loss of agro-biodiversity, uncertainty, and perceived control: A comparative risk perception study in Austria and China. *Risk Analysis*, *26*(2), 455-470.

Schosler, H., de Boer, J. & Boersema, J. J. (2013). The organic food philosophy: A qualitative exploration of the practices, values, and beliefs of Dutch organic consumers within a cultural-historical frame. *Journal of Agricultural & Environmental Ethics*, *26*, 439-460.

Scudamore, K. A., Guy, R. C. E., Kelleher, B., Macdonald, S. J. (2008). Fate of the fusarium mycotoxins, deoxynivalenol, nivalenol and zearalenone, during extrusion of wholemeal wheat grain. *Food Additives and Contaminants*, *25*(3), 331-337.

Sharma, A., Gupta, M., & Shanker, A. (2008). Fenvalerate residue level and dissipation in tea and in its infusion. *Food Additives and Contaminants*, *25*(1), 97-104.

Siipi, H. (2013). Is natural food healthy? *Journal of Agricultural & Environmental Ethics*, *26*, 797-812.

Silva, R., Martinez, M. L., Hesp, P. A., Catalan, P., Osorio, A. F., Martell, R., Fossati, M., da Silva, G. M., Marino-Tapia, I., Pereira, P., Cienguegos, R., Klein, A. & Govaere, G. (2014). Present and future challenges of coastal erosion in Latin America. *Journal of Coastal Research*, *71*, 1-16.

Stefani, F. O. P. & Hamelin, R. C. (2010). Current state of genetically modified plant impact on target and non-target fungi. *Environmental Reviews*, *18*, 441-475.

Storm, I. D., Rasmussen, P. H., Strobel, B. W. & Hansen, H. C. B. (2008). Ergot alkaloids in rye flour determined by solid-phase cation-exchange and high-pressure liquid chromatography with fluorescence detection. *Food Additives and Contaminants*, 25(3), 338-346.

Stuart, D. & Woroosz, M. R. (2013). The myth of efficiency: Technology and ethics in industrial   food production. *Journal of Agricultural & Environmental Ethics*, *26*, 231-256.

Van Haperen, P. F., Gremmen, B. & Jacobs, J. (2012). Reconstruction of the ethical debate on naturalness in discussions about plant-biotechnology. *Journal of Agricultural & Environmental Ethics, 25,* 797-812.

Voigt, K. A. (2011). Pigs in the backyard or the barnyard: Removing zoning impediments to urban agriculture. *Environmental Affairs, 38,* 537-566.

Wu, F. & Guclu, H. (2013). Global maize trade and food security: Implications from a social network model. *Risk Analysis, 33*(12), 2168-2178.

Yu, H., Huang, G. H., Zhang, X. D. & Li, Y. (2010). Inhibitory effects of organic acids on bacteria growth during food waste composting. *Compost Science & Utilization, 18*(1), 55-63.

Chapter 14

GMO INVASION

Overview

"GMO Invasion" elaborates on the history of genetically modified organisms (GMOs) and genetically modified (GM) crops, GMO regulations and policies, health, environmental and economic hazards. Monsanto's advertisement campaign in Britain and "Operation Cremate Monsanto" are further described in full.

***

Despite controversies and debate surrounding the introduction of GMOs, the agricultural biotechnology has spread worldwide since its initial induction on the United States (U.S.) market (Pizella et al., 2012).

14.1 Quick Fix Solution

In the root of agriculture since the World War II has been the vision of a quick-fix, miracle solution, designed to solve world hunger. The quick-fix movement in agriculture began with the application of synthetic pesticides, created to protect cultivated crops, followed by the development of GM insecticide and pesticide-resistant plants. These miracle solutions that "have been developed by certain industrial giants who have important economic and financial interests invested in them" have also contributed to the "huge delay" in the establishment of "sustainable agriculture." An example of a "miracle solution" is the GMO (Leo & Pintureau, 2013). GMO advocacy initiatives refer to GMOs as "weapons" against world hunger, intended to assist in feeding the ever growing human population (Toft, 2012).

14.2 Selective Breeding

Agricultural crops fall into two major categories, traditional and modern. Traditional varieties are genetically diverse and result from selective breeding conducted by the farmer. Modern varieties result from scientific breeding. They are genetically unique and generate improved yield (Schmidt & Wei, 2006). GMO has been equated to the centuries-old process known as selective breeding. Selective breeding involves the enhancement of certain genes already existing within a particular species or the introduction of traits via hybridization of similar species. It has been argued that GM techniques have an advantage over selective breeding as they ensure faster and more efficient development of desirable genes. In order to ensure faster and more effective gene transfer, GM methods involve the insertion of genes from one species to a fundamentally genetically different species (Aslaksen et al., 2006). The aim is to replicate the target genetic characteristics of the original species in the receiving organism. The practice involving the integration of genes from diverse organisms is referred to as recombinant DNA technology and the resulting organism is said to be "genetically engineered", "transgenic" or "genetically modified" (Bongyu et al., 2009). In other words, geoengineering or biotechnology is designed to introduce genes from one species (i.e., plant; animal; microorganism) into another thus producing foods containing desirable for the producer or consumer traits. The resulting product is referred to as genetically modified (GM) (Magana-Gomez & Calderon de la Barca, 2008). The "trial and error process of selective breeding" is avoided when genetic engineering is used (Bongyu et al., 2009).

14.3 What is GMO?

The European Union (EU) defined GMO as, "...an organism, with the exception of human beings, in which the genetic material has been altered in a way that does not occur naturally by mating and/or natural recombination" (Van Haperen et al., 2012). The main issue in the debate over GMOs as it pertains to global justice is the claim that GMOs are a necessity if we are to feed the world. Nobel laureate Normal Borlaug stated in the Wall Street Journal in 2000: "We need Biotech to Feed the World" (Toft, 2012). Biotechnology was proposed in the 1990s as a way to increase agricultural production and provide a solution for the depletion of agro-biodiversity (Van Haperen et al., 2012). Biotechnology is defined by the Convention on Biological Diversity as "any technological application that uses biological systems, living organisms, or derivatives thereof, to make or modify products or processes for specific use."

Genetic modification was discovered in the 1970 (Bongyu et al., 2009). Genetic engineering began in the 1980s when a Bacillus thuringiensis (Bt) gene was inserted in tobacco upon cloning it (Brana et al., 2012). Produced via biotechnology, GMOs in agriculture were first commercially introduced in the U.S. in the 1990s (Pizella et al., 2012). GM crops were planted on 145 million acres worldwide in 2002 (Bongyu et al., 2009). 143 million hectares in 23 countries of GM crops were planted between 1996 and 2007 (Magana-Gomez & Calderon de la Barca, 2008). GM crops were planted in 25 countries on 134 million hectares which represented 8.7% of worldwide arable land in 2009 (Stefani & Hamelin, 2010). The U.S., Argentina, China, Brazil, Canada and India are the largest producers of GM crops, accounting for over 90% of the world's GM production. The U.S. alone produces more than 50% of global GM crops (Magana-Gomez & Calderon de la Barca, 2008). Cotton, soybean, corn and canola as well as crops with herbicides and insecticides properties are the most widely used GMOs (Pizella et al., 2012). The major GM crop in 2007 was the soybean, accounting for 51% of global GM area followed by maize (31%), cotton (13%) and canola (5%) (Magana-Gomez & Calderon de la Barca, 2008). 70% of soybean crops worldwide were planted with GM soybean crops, 48% with GM cotton, 24% with GM maize and 20% with GM canola crops in 2009. Bt crops occupied 200 million hectars since 1996 and "have virtually replaced traditional cultivars in areas such as the U.S., South Africa, China and India. The most widespread GM woody species include the Populus, Pinus, Picea, Liquidambar and Eucalyptus (Stefani & Hamelin, 2010). Bananas containing vaccines against infectious diseases such as hepatitis B, fish that grows rapidly, nut and fruit trees that yield much earlier and plants that yield plastics with specific properties are "on the horizon."

GMO business companies and some scientists suggest that GM crops are the only means to ending world hunger in the 21st century. They argue that GMOs advantages of being cold, drought, disease, and pest resistant, herbicide tolerant, cheaper and more nutritional are incentive to large-scale GM crops production to "meet the skyrocketing demands" (Bongyu et al., 2009). GM crops are further characterized by their properties of high productivity (Leo & Pintureau, 2013). However, greater food output does not necessarily translate into increased food supply to those most in need of nutrition. Food production as opposed to food demand and health is the main factor that trigger genetic modification of agricultural crops. GM production sectors insist on increasing food production instead, a phenomenon known as the "technology-push." GM seeds are manufactured within industry sectors producing both GM seeds and herbicides. Public concerns arise when emerging technology generates great profit, yet is scientifically controversial in regards to its environmental and public health effects (Aslaksen et al., 2006). Some of the main concerns include its impact on human and environmental health (Brana et al., 2012), consumer rights and trade relationships involving GMO use (Pizella et al., 2012). Scientists disagree as to whether the modification of crops is ethical and whether GM crops are beneficial or detrimental to the environment (Gregorowius et al., 2012). Opponents of genetic modification argued that such biotechnologies would contribute to further environmental degradation, loss of agro-biodiversity, increased human health risk and agricultural appropriation (Van Haperen et al., 2012).

14.4 Environmental Risks

The environmental hazards related to GM plants and trees are nearly similar. Common risks include vertical gene flow, gene escape, horizontal gene transfer and seed dispersal (Stefani & Hamelin, 2010). Superweeds are herbicide-tolerant plants that emerge as a result of gene migration from transgenic to wild plants. GMO seeds could be released into the environment, replacing local plant species due to GMOs genetic advantages (Gregorowius et al., 2012). The transfer of genetic organisms to native plants via routes such as pollination, microbial transfer and dispersal is known as gene flow (Bongyu et al., 2009). A study, conducted by Eric Quist and Ignacio Chapela and published in *Nature* in 2001, demonstrated that five out of seven native maize species in Mexico were contaminated with transgenes. The study endured great criticism which culminated in *Nature* issuing editorial note on April 11, 2002, stating that "the evidence is not sufficient to justify the publication of the original paper." According to the authors, "withdrawing a published article is a surprising response to a situation of controversy, which normally will find its resolution through subsequent publication of new results and open debate." Other studies have also demonstrated the herbicide-resistance gene flow from transgenic plants to wild crops. For example, a harmful to agricultural crops weed called rigid ryegrass, has been shown to exhibit glyphosate tolerance properties. A concern has been therefore expressed that the use of more toxic herbicides may be necessary provided glyphosate resistance weeds disperse (Aslaksen et al., 2006).

In more general terms, the use of herbicide-resistant GM crops may necessitate an increased use of herbicides (Gregorowius et al., 2012). Furthermore, it is possible that a genetic modified species may become an invasive species (Aslaksen et al., 2006).

GM plants and trees are typically spread as monocultures over grand land areas and are specifically designed to generate improved yield while allowing for the simplification of agricultural management. Such conditions could contribute to environmental risks and stress such as the impact of GMOs on non-target species (Stefani & Hamelin, 2010). A third concern therefore includes the adverse effects of GM crops on non-target organisms such as pollinators (Bongyu et al., 2009). Toxic substances produced by GM plants may enter the food chain and affect non-target organisms (Gregorowius et al., 2012). One study demonstrated the detrimental impact of Bt crops on monarch butterfly larvae upon ingestion of Bt pollen. A different study purports that such impact is "negligible." Ladybird larvae which is beneficial in managing pernicious insects had also been adversely impacted by GM crops, particularly, corn. According to a literature review (Aslaksen et al., 2006), fungi are valuable to the environment organisms as they are responsible for decomposition, soil formation, fertility and for shaping plant populations dynamics and structures. The review analyzed 149 studies in order to demonstrate the impact of GMOs on target (84 studies) and non-target fungi (35 studies). Results from the study revealed that GM plants and trees with anti-fungal properties are generally successful in decreasing the impact of fungal pathogens on GM crops. Five studies revealed detrimental impacts of GMOs on clone-specific and soil-specific non-target fungi. The majority of reviewed studies were conducted under controlled as opposed to field conditions. Extrapolation of controlled-field findings to field conditions is therefore not appropriate as environmental variables have not been factored within the study and the results (Stefani &

Hamelin, 2010).

It has also been suggested that Bt corn can excrete toxins into the soil via its roots, impacting the soil microfauna which plays a significant role in decomposing organic matter in the soil. GM crops could potentially contaminate native crops with GMO via pollen flow and contribute to reduced genetic diversity in native crops (Aslaksen et al., 2006). GM crops incessantly produce toxic molecules at large doses, acting to obliterate specific bio-agressors. Such toxic molecules attach to soil structures and remain active in soils for more than 284 days. GM crops are moreover incompatible with other crops, posing additional environmental risks (Leo & Pintureau, 2013). Modification of plants is also morally unacceptable as it deviates from natural order by undermining their dignity or integrity (Gregorowius et al., 2012). Lastly (Aslaksen et al., 2006), results from studies on animal species fed with GM foods reveal that "organelles and other subcellular structures are clearly affected" (Magana-Gomez & Calderon de la Barca, 2008). Environmental risk concerns remain despite the large amount of research conducted relative to GMO risk assessment. It has been argued that it will take at least 30 years to assess the environmental risks pertaining to GMOs release (Aslaksen et al., 2006).

14.5 Human Health Risks

Scientists, governmental officials, and public representatives believe that GMOs contribute to human health risks including lower nutritional values, possible reduced concentrations of phytoestrogens which protect against cancer and cardiovascular conditions, allergic reactions resulting from the insertion of new genes into crops and animals, antibiotic resistance and (Bongyu et al., 2009) direct absorption of genetic particles (Aslaksen et al., 2006). Children in the U.S. and Europe have exhibited severe allergic reactions to certain foods such as peanuts. Foodborne disease which is exponentially growing worldwide, is another human health risk argued to result from ingestion of GMOs. 30% of people living in industrialized countries have experienced foodborne illness. 76 million people are diagnosed with a foodborne disease each year in the U.S., resulting in 5000 deaths and 325,000 hospital admissions. 130 million people are diagnosed with a foodborne disease in Central Asia and Europe each year. The intricate GMOs production systems further add to the potential for contamination of produce (Bongyu et al., 2009). "The scientific priority is to contribute to the improvement of human and animal health or natural resource management without compromising public safety" (Magana-Gomez & Calderon de la Barca, 2008).

14.6 Economic Risks

One economic risk related to GMO concerns the possibility that the global food supply will be increasingly controlled by only a few large corporations. Another economic risk focuses on the transference of ownership rights over genetic resources to private companies (Aslaksen et al., 2006). Developed countries "have monopolized the GMO technology through strict patents and licensing agreements leading to marginalization of developing countries" (Bongyu et al., 2009). Additional related risks involve the presence of GM crops containing terminator seeds which inherently do not produce fertile seeds, farmers being obligated to buy GMO seeds as opposed to producing their own and farmers being vulnerable to litigation provided their native crops are accidentally contaminated by GMOs (Aslaksen et al., 2006). Other economic problems relative to GM crops include the high cost of seeds in developing countries, the possibility of contaminating conventional crops as a result of gene transfer and political problems such as the monopolization of seeds by a few companies (Leo & Pintureau, 2013).

14.7 GMO Regulation, Policies and Procedures

The scientific debate and disagreements regarding GMOs' impact on environmental and human health safety has resulted in significant confusion as well as suspicion in society at large. Similar to all technologies, GMOs could be either used to benefit those in need or to profit those political, technological, economic sectors holding power. The transfer of genes between organisms contributes to uncertainties in terms of the ways genes would interact and the environmental and human health impact. It is therefore imperative that strict regulations ensuring human and environmental safety accompany the release of GMOs (Bongyu et al., 2009). It is argued that regulatory policies pertaining to GMOs use need to take into consideration environmental, cultural, ethical, economic as well as social aspects (Pizella et al., 2012). Public participation and the right to informed consent must be furthermore respected as part of the decision-making process (Bongyu et al., 2009).

Governments are responsible for the regulation of the agricultural sector, including food production and distribution (Bongyu et al., 2009). GMOs-pertaining laws and regulations have been established by governments as a means to address national and international concerns related to GMOs release (Brana et al., 2012). International regulations of GMOs fall into two main categories: regulations promoting free trade and regulations concerned with biosafety of GMO plants (Toft, 2012). The Cartagena protocol has been established to ensure international biosafety relative to the use of GMOs (Aslaksen et al., 2006). "Historically, industry has proven unreliable at self-compliance of existing safety regulations" (Bongyu et al., 2009).

The establishment of standardized guidelines to assess the safety of foods derived from biotech crops was deemed necessary as per "the introduction of recombinant DNA technology in plant breeding." No standardized assessment to test safety of GM foods exists to date (Magana-Gomez & Calderon de la Barca, 2008). Various countries have devised different GMO-related policies (Aslaksen et al., 2006).

14.7.1 "Permissive" vs. "Restrictive" Regulatory Approach

In 1993, the Organisation for Economic Co-operation and Development (OECD) proposed substantial equivalence (a comparison between conventional and GM foods) as the basis for ensuring safety of biotech products (Magana-Gomez & Calderon de la Barca, 2008). Agricultural biotechnology products (i.e., GMOs) in Canada and the U.S. are considered to be "innocent until proven guilty." As long as GMOs production companies can demonstrate that GMO products are "substantially equivalent" to their conventional counterparts, no extensive risk assessments are required (Clapp, 2005). It has been argued that substantial equivalence is not a safety assessment due to its inability to detect "unintended effects" (Magana-Gomez & Calderon de la Barca, 2008). The European Union's (EU's) approach to agricultural biotechnology products is "much more precautionary". "Rigorous" risk assessments are undertaken prior to the approval of GMO products. In that sense, agricultural biotechnology products are perceived as "guilty until proven innocent" (Clapp, 2005). While the U.S. regulations regarding GMOs are considered as "permissive", the EU regulations are regarded as more "restrictive" (Toft, 2012). The U.S. views the EU regulatory system as being overly "emotional" and "not scientifically based" (Clapp, 2005).

14.7.2 U.S. Regulation Agencies

Three main agencies are responsible for regulating GM crops in the U.S. The Environmental Protection Agency (EPA) determines the environmental safety of GM crops, the Food and Drug Administration (FDA) is responsible for assessing whether the food is safe for consumption and the U.S. Department of Agriculture (USDA) shares the risk assessment responsibilities of GM crops and evaluates whether the plants are safe to grow. A particular FDA review is mandated only if GM crops contain substantially different substances, such as toxins, compared to the ones found in conventional crops, include substances that can cause antibiotic resistance, allergic reactions and unanticipated genetic effects, and contain significantly different levels of nutrients. In other words (Bongyu et al., 2009), formal public risk assessment is not mandatory prior to releasing GM crops in the U.S. as long as the USDA had been notified of the intended release of GM crops. Two environmental safety reviews by USDA and by the EPA are required only when planning on releasing GM crops known to produce toxins (Aslaksen et al., 2006).

14.8 The Precautionary Principle

OECD established its first safety guidelines in regards to GMOs. By 1993, safety principles in biotechnology were primarily based on the substantial equivalence and familiarity principles which concern the degree of similarity between traditionally bred crops and GMO crops. Scientific gaps remain in the long-term assessment of environmental risks relative to GMO release. A sole reliance on the substantial equivalence and familiarity principles may not suffice in thoroughly assessing risks pertaining to GMOs. Such environmental risks may be "potentially irreversible." Improved regulatory policies incorporating the precautionary principle could therefore prove beneficial (Aslaksen et al., 2006).

The precautionary principle is used in biotechnology to manage any risks relative to GMOs. The precautionary principle posits that any biotech product that is not proven to be safe is considered unsafe and should therefore be withheld from commercial release (Bongyu et al., 2009). The precautionary principle which is part of environmental risk management was established in 1992. Debate has emerged regarding when and how to use the precautionary principle. The precautionary principle has been incorporated in EU legislation concerning GMOs. EU guidelines propose that the application of the precautionary principle corresponds to expected environmental risks. It has been recognized that risk assessment is insufficient in fully identifying all possible environmental risks. Controversies therefore exist in terms of how to interpret and subsequently incorporate the precautionary principle within decision-making processes regarding the use of biotechnology. The European Commission has proposed the integration of the precautionary principle within a RA framework. A complete RA has been suggested in relation to the release of GMOs in the environment given the environmental as well as economic risks pertaining to the application of GM technologies. It has been proposed that improvements in risk communication between various stakeholders (i.e., government; scientists; consumers; industry) could also improve risk assessment and risk management, allowing for the identification of "early warnings".

The "large economic incentives for early adoption of GMO," "the widely divergent interest and risk perceptions of stakeholder groups" and "early warnings of hazards...discounted by the interests of various stakeholders" prevent the implementation of the precautionary principle relative to genetically modified technology (Aslaksen et al., 2006). For example, large GMO producing countries, such as the U.S., Canada and Argentina, filed a complaint against the EU in 2003, arguing that the focus on the precautionary principle as part of EU regulation policy has contributed to delays in the approval of GM crops. EU was not expected to change its regulation policies regarding GMOs. The EU was rather expected to not delay GMOs approvals solely on the basis of the precautionary principle (Toft, 2012).

The precautionary principle, designed to manage

situations posing risk, has been criticized for being indistinct and for not integrating an adequate scientific database as part of the GMOs decision-making process (Bongyu et al., 2009). A precautionary approach to increasing food production does not involve the sole application of biotechnologies in agriculture. Rather, it incorporates the continual development of organic and traditional agriculture, the restoration of environmental health and the conservation of genetic crop diversity (Aslaksen et al., 2006).

14.9 Mistrust

Major points incorporated within the debate concerning regulation of GMOs include human health and environment risks, who the participants of risk analysis and decision-making are and food labeling and consumers' right to information regarding what they consume (Bongyu et al., 2009). Trust is the foundational block of social institutions. Trust generates a sense of community, predictability and ease in working together. It is "easier to influence or persuade someone who is trusting" (Folke et al., 2005). People mistrust the government, believing that the government and the devised by the government policies and regulations, surrender to the economic interests of businesses. Such fear was highlighted by the Tuskegee study where humanity was used as "guinea pigs." It has been deduced that large corporations can use the enormous profits that they generate to circumvent governmental regulations. Biotechnology is currently an enormous industry with great potential for profit. Monsanto expressed dissatisfaction when it generated a profit of $294 million on $7.5 million sales. GM companies such as Monsanto, Dupont and Novartis invest billions of dollars on the production and research of GM foods. Scientific organizations such as the Rowett Research Institute have become dependent on financial sponsorship from such companies due to governmental cuts. There is a trend of a few large corporations governing the entire GM market.

The Center for Science in the Public Interest (CSPI) recommended a strengthening of the FDA review process in order to alleviate consumers' concerns regarding GMOs safety. The Ministry of Health and Welfare had mandated risk assessment of GMOs prior to its release. Several states in Brazil have banned GMO foods while efforts to prevent the import of GMOs in Brazil have been undertaken (Bongyu et al., 2009). 35 countries worldwide comprising half of the population of the world "have rejected GM technology" (Clapp, 2005). The World Trade Organization (WTO) has precluded countries from banning GM foods despite referendums and widespread public petitions and protests (Bongyu et al., 2009). Critics of GMOs believe that the WTO which takes part in GMOs regulations is "working in favor of the rich countries and to the detriment of developing countries" (Toft, 2012). It is not the case that research on GM crops is not being conducted. It is being conducted. However, it is focused mainly on market and corporate objectives rather than on safety. It is furthermore not uncommon that only studies supporting organizations' political positions are propelled while interpretation of scientific data is skewed to "suit their needs." Consumers unaware of GMOs safety risks are most vulnerable to consuming GM crops which can possibly harm their health. It is therefore advised that such consumers are "protected against the aggressive and rich GMO companies that can use every means possible to ensure that they maximize their profit." Not all people have the required education to participate in the GMO debate. Further, not all consumers have access to the necessary resources and information to make an informed decision as to whether to consume GM foods or not. The Precautionary Adoption Process Model (PAPM) has been proposed as a way to identify obstacles to undertaking preventive actions and to shift public awareness from a place of unfamiliarity to a state of action in regards to GMOs. It is the responsibility of scientists, experts and the government to educate the public regarding biotechnology and GMO safety (Bongyu et al., 2009).

14.10 GMO Brazil

Brazil was the second largest country after the U.S. to grow genetically engineered agricultural plants in 2009, representing 16% of all GM crops grown in the world or occupying 21.4 million hectares of land. Argentina takes third place with 21.3 million hectares of planted GM crops (Brana et al., 2012). The National Commission of Biosafety (CTNBio) was established by Biosafety Law to govern the regulatory and decision-making aspects relative to the commercial release of GM crops in Brazil. Risk Analysis (RA) and Environmental Impact Assessment (EIA) are prerequisite to the commercial release of GM crops in Brazil (Pizella et al., 2012). RA consists of three components – risk assessment, risk management and risk communication (Bongyu et al., 2009).

14.10.1 GMO RA in Brazil

A criticism of RA as a means of assessing the biosafety of GM crops is its short duration and limited area scope. In other words, it does not allow for long-term assessment of GMO and for an observation of its interaction with various environmental factors. Results from studies conducted within a few Brazilian GM crop locations are extended to the entire territory of Brazil. No structures for the monitoring of GMO release in Brazil currently exist despite the fact that GM crops have been commercially released on the entire national territory with the exception of protected by law Environmental Areas and Indigenous Lands. Monitoring data pertaining to large-scale GM crop release in various countries is not available, further precluding the conduct of an efficient RA in Brazil. Systematic RA in Brazil is hindered by the lack of scientific data demonstrating the effects of GM crops on human health and the environment. Such risks are not easily predictable. Environmental risks are typically identified not until the "damage occurs." RA in Brazil further fails to consider cropping or location alternatives to GMO systems. Cultural, environmental, political and economic aspects pertaining to GMOs are not taken into consideration as part of the decision-making process. Furthermore, "biosafety and environmental gaps...are not questioned by the Commission." All biosafety studies primarily focused on comparing the agronomic production features of GM crops and conventional crops were "until now...held at the headquarters of the companies involved." "If conducted in a transparent and independent form by the competent agencies," RA "may be understood as a punctual evaluation tool of the cultivar" (Pizella et al., 2012). Collaboration between botanists, medical professionals, statisticians, social scientists, entomologists, ecologists and microbiologists is required to improve RA (Aslaksen et al., 2006).

14.10.2 RA and EIA in Brazil

It has been argued that RA and EIA are currently inefficient in detecting human health and environmental risks. It has further been postulated that even if performing to their optimal potential, RA and EIA are unable to determine GMOs biosafety risks as they are specifically designed to be used in local rather nation-wide territories. A concern has been further expressed regarding the scarce public participation in the GM crops release decision-making process. Public hearings relative to GM crops release are only held in the capital of Brazil, Brazil, and are not made mandatory (Pizella et al., 2012).

14.11 Labeling of GMO

The introduction of genetically engineered foods sparked a

debate relative to the labeling of conventional and GM foods. While proponents believe that labeling of GM foods as such is not necessary as long as GM foods are safe for consumption and have the same nutritional value as conventional foods, opponents defend the position that labeling of GM foods is imperative. Opponents believe that consumers "have a right to know" information pertinent to safety and nutritional value, but also information "about genetic modification and the kind of production method that is used, even if this information has no impact on the quality of the final product" (De Tavernier, 2012).

A number of European countries, Australia, Japan, New Zealand, South Korea and China have established labeling of GM foods as mandatory. The FDA, however, proposed that labeling of GM foods is voluntary. Such decision is based on two arguments. First, labeling of foods as GM is equivocal as GM and non-GM foods are cross-pollinated. Secondly, labeling of GM foods as such implies that non-GM foods are of greater quality compared to GM foods. The question to ask therefore is how to educate the public regarding GM foods without causing them to mistrust or fear GM foods (Bongyu et al., 2009).

14.12 Monsanto

Smith (2002: 9) states that for the majority of commercial organizations "the bottom line is profit" (O'Keeffe, 2009).

14.12.1 British Advertising Campaign

Monsanto, a U.S.-based biotech and GM foods corporation, organized British advertising campaign in 1998. The aim of the campaign was to alleviate public concerns regarding GM foods by presenting the benefits of biotechnology and more specifically, GM foods. The campaign was comprised of several full-page advertisements. Headlines of the ads included: "Food technology is a matter of opinions. Monsanto believes you

213

should hear all of them," "We believe food should be grown with less pesticides," "Food labelling. It has Monsanto's full backing," "This strawberry tastes just like strawberry," "If it weren't for science, her life expectancy would be 41 years," "Worrying about starving future generations won't feed them. Food biotechnology will." Such advertisement is considered to be an example of "Greenwash." "Greenwash" denotes the attempt of corporations to increase their market growth by posing as environmental allies, leaders in the efforts to end poverty. A Greenpeace spokesperson referred to Monsanto's ads as "Greenwash with a guilt trip," particularly concerning the heading: "Worrying about starving future generations won't feed them. Food biotechnology will." One of the ads predicted: "With the planet set to double in numbers around 2030, this heavy dependency on land can only become heavier. Soil erosion and mineral depletion will exhaust the ground. Lands such as rainforests will be forced into cultivation. Fertilizer, insecticides and herbicide use will increase globally." Several repetitive themes could be observed in the ads. One included the attestation that, "Of course Monsanto is a business. We aim to make profits, acknowledging that there are other views of biotechnology than ours." Another emerging theme depicts GM foods as diligently studied for safety with 20 governmental regulatory agencies having had approved such studies. Yet additional themes proposed that Monsanto is integrating the best of the traditional and modern world and that the reader is to be aware that there are different opinions. Monsanto discharged the advertising agency following the end of the campaign which was only partially successful. The agency had advised Monsanto to abstain from running the campaign prior to its release in the first place. Monsanto USA prohibited the use of GM foods within the company's cafeteria several months following the campaign and in the midst of controversies regarding GMOs safety taking place in Britain and the U.S. (Reynolds, 2004).

14.12.2 "Operation Cremate Monsanto"

It is common that protest movements such as the ending of slavery and colonial rule either do not succeed or take decades to make a difference. A national protest movement called "Operation Cremate Monsanto" was initiated in India upon the approval of the first GM crop, Bt-cotton, promised to decrease pesticide use and control the number one predator of cotton, the bollworm. Monsanto's Bt-cotton was approved by the government of India in March, 2002, yet Bt-crops were grown by farmers even before Bt hybrids official approval. Over 50% of GM cotton in India is derived from illegal sources. Approximately 35% of cotton seeds sold in India are transgenic of which 9% are legal and 25% - illegal. 7,907,200 acres of Bt hybrids were planted by 2006. One of the claims of "Operation Cremate Monsanto" movement concerned the so-called terminator genes which are believed to produce sterile plants. The so-called suicide seeds could escape to the environment and contaminate wild plants with sterile genes. Such claim, according to the author, is incorrect as the terminator gene technology called gene use restriction technology (GURT) is not owned by Monsanto and has not been released for commercial use. A different claim was proposed, attesting that the high suicide rate amongst farmers in India resulted from the debt associated with the use of terminator seeds. Terminator gene crops, it was argued, required farmers to continue purchasing new seeds every season while generating low return thus increasing farmers' debt. The protest movement proposed that 40,000 Indian farmers have committed suicide due to the use of GM technology. On the opposite, farmers in India adopt Bt-cotton as it reduces their dependence on pesticides, facilitates insect control and therefore increases their income. A final claim is that thousands of sheep have died as a result of consuming the leaves of Bt-cotton crops. Pesticide companies are in opposition to Bt-cotton as it contributes to reduced use of their products. The movement failed to prevent the introduction of GM crops in India primarily due to the divergent opinions between protesters and farmers and between protesters' ideology and empiricism (Herring, 2006).

Discussion

It could be concluded that humanity and Earth are being poisoned by governments and industries without any consequence for committed crimes, yet a farmer whose crops are accidentally poisoned by GMOs is to be sued. Such a reality depicts an ideal example of how power is gained.

Furthermore, governmental regulations of GMOs in countries such as the U.S. and Canada are based upon the principle of substantial equivalence which literally translates into GMOs being "innocent until proven guilty." Yet, the article exposing the contamination properties of GMOs was retracted from "Nature," which is a clear indication that regulations are the deceit beneath which crimes take place. GMO research is moreover literally owned by GMO companies and not publicly disseminated. The above two examples highlight the deliberate difficulty of proving that GMOs are "guilty." Such created problem translates into GM companies continuing to generate profit. Monsanto even confirms that it is a business first and foremost. It even has a plan of how it will continue profiting through its practices, taking into consideration the created continuation of environmental deterioration. Is it then surprising that pesticides are owned by GMO companies? What clearer example could there be that problems are intentionately created to generate profit?

Lastly, humanity is gaining awareness of the toxic effects of GMOs, which, according to Sosé, is the first step to implementing efficient solutions to existing problems.

References

Aslaksen, I., Natvig, B. & Nordal, I. (2006). Environmental risk and the precautionary principle: "Late Lessons from Early Warnings" applied to genetically modified plants. *Journal of Risk Research*, 9(3), 205-224.

Bongyu, M., Billingsley, G., Younis, M. & Nwagwu, E. (2009). Genetically modified foods and public health debate: designing programs to mitigate risks. Public Administration & Management, 13(3), 191-217.

Brana, G. M. R., Miranda-Vilela, A. L. & Grisolia, C. K. (2012). A study of how experts and non-experts make decisions on releasing genetically modified plants. *Journal of Agricultural & Environmental Ethics*, 25, 675-685.

Clapp, J. (2005). The political economy of food aid in an era of agricultural biotechnology. *Global Governance, 11*, 467-485.

De Tavernier, J. (2012). Food citizenship: Is there a duty for responsible consumption? *Journal of Agricultural & Environmental Ethics, 25*, 895-907.

Folke, C., Hahn, T., Olsson, P. & Norberg, J. (2005). Adaptive governance of social-ecological systems. *Annual Review of Environment & Resources, 30*, 441-473.

Gregorowius, D., Lindemann-Mathies, P. & Huppenbauer, M. (2012). Ethical discourse on the use of genetically modified crops: A review of academic publications in the fields of ecology and environmental ethics. *Journal of Agricultural & Environmental Ethics, 25*, 265-293.

Herring, R. J. (2006). Why did "Operation Cremate Monsanto" fail? Science and class in India's great terminator-technology hoax. *Critical Asian Studies, 38*(4), 467-493.

Leo, C. & Pintureau B. (2013). Crop protection between sciences, ethics and societies: From quick-fix ideal to multiple partial solutions. *Journal of Agricultural & Environmental Ethics, 26*, 207-230.

Magana-Gomez, J. & Calderon de la Barca, A. M. (2008). Risk assessment of genetically modified crops for nutrition and health. *Nutrition Reviews, 67*(1), 1-16.

O'Keeffe, J. (2009). Sustaining river ecosystems: Balancing use and protection. *Progress in Physical Geography, 33*(3), 339-357.

Pizella, D. G. & Souza, M. P. (2012). Brazilian GMO regulation: Does it have an environmental approach? *Journal of Environmental Assessment Policy and Management, 14*(2), 1-16.

Reynolds, M. (2004). How does Monsanto do it? An ethnographic case study of an advertising campaign. *Text, 24*(3), 329-352.

Schmidt, M. R. & Wei, W. (2006). Loss of agro-biodiversity, uncertainty, and perceived control: A comparative risk perception study in Austria and China. *Risk Analysis, 26*(2), 455-470.

Stefani, F. O. P. & Hamelin, R. C. (2010). Current state of genetically modified plant impact on target and non-target fungi. *Environmental Reviews, 18*, 441-475.

Toft, K. H. (2012). GMOs and global justice: Applying global justice theory to the case of genetically modified crops and food. *Journal of Agricultural & Environmental Ethics*, 25, 223-237.

Van Haperen, P. F., Gremmen, B. & Jacobs, J. (2012). Reconstruction of the ethical debate on naturalness in discussions about plant-biotechnology. *Journal of Agricultural & Environmental Ethics, 25*, 797-812.

Chapter 15

ANIMAL MUTILATION

Overview

"Animal Mutilation" elaborates on the health risks pertaining to meat consumption and livestock welfare. Policies and environmental hazards pertaining to livestock farming are discussed. Solution to the meat demand crisis is proposed.

15.1 Infection

"A number of infectious diseases have emerged as threats to humans and wildlife." A great number of infectious diseases threatening wildlife also endanger human health. 60% of all infectious to human beings pathogens and 75% of all emerging infectious agents are zoonotic. 60% of infectious disease in human beings results from fungi and bacteria. Contact between human beings and wildlife has contributed to disease outbreaks in human populations in the past (Hopkins & Nunn, 2007).

Trichinellosis is a human disease caused by the consumption of undercooked meat contaminated with the parasite Trichinella nativa (T. nativa). T. nativa is particularly adapted to the arctic and subarctic regions of North America. 1.5% of black bears in Canada are infected with T. nativa. except for the Kootenay region of British Columbia where 12% of black bears were found to be infected. 27% of black bears in Alaska and 5-10% of black bears in Montana are infected with the parasite. Diarrhea, abdominal pain, nausea, vomiting, fever, muscle weakness, rash, edema, leukocytosis and eosinophilia are the symptoms experienced by human beings upon infection with T. nativa. Gastrointestinal impairments may last up to 3 weeks. Antiparasitic drug treatment has been found effective in treating trichinellosis (Schellenberg et al., 2003).

15.2 Animal Welfare

Animal ethics refers to the means of caring, housing, breeding and feeding animals (Webb, 2013). Animal husbandry and slaughtering animals for consumption have been practiced by human beings for thousands of years. Labor, food production and companionship have been the main reasons for raising animals. Sacrificing an animal's life for ritual purposes has also been prominent historically for a large proportion of the human population. According to the Bible (Genesis 1:29), the first human beings were herbivores and not until later on were they permitted to consume animal flesh (Genesis 9:2-4). The Bible and human history further depict human beings as being "dominant over other life forms." Animal welfare regulations were non-existing among the majority of societies during most of this time period (Zivotofsky, 2012).

The growing demand for food has shifted the structure of animal production systems. Traditional farms were extensive, encompassing large areas where animals could roam and where livestock would be treated kindly. Many modern farms, on the other hand, were designed as more intensive factory farms, owned by influential corporations that failed to treat animals with kindness. Intensive animal production systems have been equated to confinement systems "disastrous for animal welfare." Such view is shared by advocates of animal rights movements on limited data. Animal rights advocates further criticize intensive production systems for contributing to anthropogenic greenhouse gas (GHG), environmental degradation, compromised animal welfare and human health. Intensive animal production systems are prominent only in certain countries and are designed for particular animals such as poultry and pigs. Intensive fattening of animals is a practice also spread only in certain parts of the world resulting from consumers' preference for a particular type, consistently and quality of meat products (Webb, 2013).

15.2.1 Jewish Religious Slaughter (Shechita)

From its inception, Judaism had established animal welfare regulations as Jewish religion requires that human beings act with compassion towards animals. One Jewish dietary law prohibits the consumption of meat from an animal that has died in any manner different than shechita. Shechita is a particular means of slaughtering an animal that ensures the least amount of pain experienced by the animal, leading to unconsciousness within seconds. Extensive religious and practical training culminating in the obtaining of a license is the requirement to perform shechita. Shechita practitioners or shichets are widely revered within their communities. Jewish law prohibits the consumption of animals slaughtered via shechita if animals are found to suffer from health conditions which would pass a veterinary inspection. Jewish law also prohibits the consumption of particular parts of the slaughtered animal such as the blood, sciatic nerve and particular fats.

There are two main types of legislation in regards to meat production. The first legislation requires electrical, gaseous or mechanical stunning of the animal prior to slaughter. Such law essentially prohibits shechita as shechita prohibits animal stunning. The second law requires that shechita animal products are labeled as such on the consumer's market as shechita is a "less acceptable" method of slaughter. It is argued that while the EU would like to ban shechita, it permits bullfighting and hunting which is an example of slaughter without stunning. 10,000 animals die from hunting and bullfighting without stunning on a yearly basis, yet the shechita method is being banned. Thousands of animals are killed without stunning, yet only shechita is stigmatized with a label (Zivotofsky, 2012).

15.2.2 Cow Welfare

There are a number of viewpoints regarding the welfare of animals and particularly their longevity. One view proposes that cows are sentient beings, capable of experiencing pain and deserving of respect for their life (i.e., considering animals' feelings; ensuring the experience of positive feelings). A different view perceives life as an inherent right of living beings. Therefore, the life of dairy cows is to be protected (Bruijnis et al., 2013).

15.2.2.1 Cow Foot Disorder

It is feasible to maintain a profitable farm without needing to adapt farming methods to the needs of dairy cows. The diverging interests of farmers and dairy cows result in "animal welfare problems like behavioral and social restriction, overcrowding, and production-related diseases specific to the systems the animals are kept in." Foot disorders, resulting from diverging interests of farmers and animals, are the most prominent health impairment affecting dairy cows. Lameness is a consequence of the development of foot disorders. The development of foot disorders in cows contribute to their premature culling.

    In the past ten years, the average dairy milk production in the Netherlands has increased from 7,500 l per cow to 8,000 l. During the same period, the number of farms decreased while the number of cows per farm increased from 60 to 75 cows on average per farm. An economically and labor efficient housing system for dairy cows is the cubicle housing system. The cubicle housing system is composed of cubicles which allow cows to lie and walking corridors mostly made of concrete floors. About 80% of cows have one or more foot disorders. One-third of cows with foot disorders develop lameness each year. Foot disorders are of long duration, causing great pain to cows. Difficulty walking, standing up and lying down result from such intense pain. A decline in milk production and reduced reproductive performance are also health consequences of foot disorders in cows. There is a greater probability of culling or prematurely slaughtering cows that have developed clinical foot disorders and lameness. Instead of being slaughtered for human consumption, however, lame cows are euthanized and disposed of due to food safety concerns. A dairy cow can live up to 20 years. Currently, the average lifespan of a dairy cow is 6 years. The Farm Animal Welfare Council (FAWC) has postulated that a 2-year increase of the lifespan of dairy cows is feasible (Bruijnis et al., 2013).

15.2.3 Piglet Castration

In 2005, more than 103 million pigs were produced in the United States (U.S.) (Sapkota et al., 2007). Piglets castration without anesthesia is a common practice in organic and conventional farming, aimed at preventing boar taint or the displeasing flavor and odor in pork. Piglet castration without pain relief is legal in the first seven days of the piglet's life. The European Union (EU) banned the practice specifically in organic agriculture in 2012 as a result of rising animal welfare concerns. Alternative methods such as castration with anesthesia and analgesics, immunocastration (vaccination preventing testicular development) and fattening of boars have been proposed. The drawback of castration with pain relief is the high expense of the procedure while the weakness of fattening is the possibility that the entire boar may develop boar taint (Heid & Hamm, 2012).

15.2.4 Organic Livestock Farming

Naturalness could be defined as leading an autonomous, self-regulated life characterized by inherent life processes. Naturalness is at the heart of organic farming. Natural behavior, reproduction and growth characterize natural living. Therefore, animal welfare is related to its habitat. Organic animal farming is characterized by "the provision of natural living conditions for animals." In truth, however, the concept of naturalness is not always attended to in organic animal farming. Premature weaning, separation of the calf from the cow and dehorning are widespread practices in organic animal farming. EU organic farming regulations ensure proper medical treatment, prohibit tethering and confining calves after the age of 1 week. EU regulations further require 1.5 m² for indoor and 1.1 m² for outdoor space per calf. Access to outdoor areas is required except for when indoor housing systems allow for movement during the winter and when pasture is made possible during grazing. Calves are to be fed with preferably maternal milk for at least 3 months. In sum, European organic farming regulations only ensure weaning age of calves of at least 3 months old, access to outdoor areas and provision of natural feed. The age of separation between cows and calves, the lack of mandatory natural suckling, the limited space per calf and outdoor access are not addressed by EU legislation (Vetouli et al., 2012).

The prime goal of organic livestock farming, in comparison with non-organic livestock farming, is the improvement of animal welfare and health. The three basic premises of animal welfare include feeling and functioning (i.e., health) well and leading a natural life. The International Federation of Organic Agricultural Movement (IFOAM) established four basic ethical principles designed to guide organic farming: health, ecology, fairness, and care. Organic farming has historically understood the principle of health as: "healthy soil gives healthy plants that feed healthy animals and healthy humans, who then feed the soil (with, among others, manure from healthy animals)." IFOAM adds: "Organic Agriculture should sustain and enhance the health of soil, plant, animal, human and planet as one and indivisible." One way of viewing animal rights includes the understanding that human beings do not own animals; animals have the right to life and to security from assault on their physical existence. One perspective on animal agriculture proposes that human and animal beings have lived together for thousands of years. According to the same perspective, as long as human beings attend to the needs of animals, including allowing them naturalness, it is morally appropriate for the human being to use animal products and take the life of animals (Vaarst & Alroe, 2011).

15.3 Contamination

World War II set the stage for technological advancements in food processing systems. Through the implementation of technological tools, food production companies focused on increasing economic efficiency, food production and profitability while decreasing investments in labor. Technological advancements in industrial food production have contributed to an increase in large-scale foodborne disease outbreaks derived from pathogens such as Escherichia coli (E. coli) and Salmonella (Stuart & Woroosz, 2013).

Nontyphoidal salmonellosis is the second leading zoonotic disease in the EU, causing gastroenteritis in human beings. Gastroenteritis is characterized by symptoms of nausea, diarrhea, headache, abdominal pain. Diarrhea may evolve into more serious illness. Pork, eggs and poultry contamination are major routes for the spread of Salmonella. (Bollaerts et al., 2010). "Industrial actors accumulate wealth while more consumers are subject to illness" (Stuart & Woroosz, 2013).

Currently, the top four slaughterhouses in the U.S. control about 84% of the beef market and process approximately 94,000 cattle per day. Beef production facilities mix meat from a variety of sources. One study found that ground beef was sourced from 11 different companies and mixed in 80-pound packages prior to reaching the grocers. Such process could substantially increase the risk of meat contamination while reducing meat traceability. There were 9,824 foodborne illnesses associated with beef consumption between 1998 and 2007. A different source of contamination is the slaughter facility. Large beef suppliers tend to refuse to sell beef meat to plants that examine the incoming meat products for pathogens, causing plants to mix untested meat products with their existing products. The Food Safety and Inspection Service (FSIS) found greater ground beef contamination in smaller plants compared to large-scale beef processing plants.

Government agencies and meat processing companies were obliged to respond to food safety concerns. The Hazard Analysis and Critical Control Points (HACCP) mandated that red meat production facilities monitor and regulate their own sanitation standard operating plans (SSOPs). "In meetings behind closed doors, powerful industrial actors had great capacity to influence state policy-makers," demonstrating "how rule-making benefits certain groups more than others." "Players in the food industry maintain great influence over government." Technological interventions to systemic problems were designed as a response to HACCP regulations. Technological fixes, do not, however, address fundamental challenges relative to contamination. Preventative actions to reduce contamination at its source were not undertaken by companies. Ionizing radiation is designed to kill microorganisms via damaging bacteria's DNA. Carcass cleaning with acid solutions and decontamination with acids and ammonia are two of the implemented quick fix intervention. Some of the technological fixes, however, were found to result in the further spread of pathogenic bacteria rather than in its elimination. Technological fixes failing to address systemic problems are indicative of the "prioritizing of profitability over consumer well-being." There is a need that the current food production systems are reevaluated (Stuart & Woroosz, 2013).

15.4 Health Risks

15.4.1 Heterocyclic Aromatic Amines (HAAs)

Heterocyclic aromatic amines (HAAs) are formed during the heating process of organic compounds containing nitrogenous substances such as protein. More than 20 HAAs have been identified as mutagens, capable of inducing mutations and promoting tumors. HAAs are mutagenic for bacteria and mammalian cell systems. HAAs could result in "chromosomal aberrations and sister chromatid exchanges in cultured cells." Out of the nine HAAs tested by the IARC, eight were possible class 2B human carcinogens and one was a probable class 2A human carcinogen. Pan-frying, grilling/barbecuing and oven-broiling of meat produce the highest concentrations of HAAs. Boiling, deep-frying, oven roasting and charcoal grilling produce reduced levels of HAAs. There is a delay of several years between exposure to carcinogenic agents and disease outcome as cancer takes long time to develop (Alaejos et al., 2008).

15.4.2 Cancer

15.4.2.1 Ovarian Cancer

Ovarian cancer is the eight most widespread cancer and the fifth cause of cancer death in women in the U.S. Environmental factors have been identified as responsible for the development of ovarian cancer in women. Oral contraceptive use has been linked to decrease risk while menopausal hormone therapy (MHT) and high fat intake have been associated with increased risk of ovarian cancer. Oral contraceptive use, high fat intake and MHT affect hormonal levels either promoting or preventing ovarian carcinogenesis. Study findings reveal that animal fat intake, yet not plant fat intake, was positively associated to an increase in ovarian cancer risk in women in the U.S. (Blank et al., 2012).

15.4.2.2 Cardiovascular Disease (CVD) and Colon Cancer

Red meat is defined as lamb, pork and beef most of the time (Eichholzer & Bisig, 2000). High red meat consumption has been correlated to increased risk of cardiovascular disease (CVD) and colon cancer. CDC is one of the leading causes of mortality and morbidity in the U.S. (Del Bo et al., 2013).

15.4.2.3 Stomach, Colorectum & Pancreas Cancer

Scientific evidence linking meat consumption with stomach, colorectum and pancreas cancer risk is increasing (Taylor et al., 2007).

15.4.2.4 Breast Cancer

A United Kingdom (UK) Women's Cohort Study investigated the association between meat consumption and breast cancer risk in 35,372 women between 1995 and 1998. High total meat consumptions were association with an increase in premenopausal breast cancer with processed meat showing the strongest hazard ratio (HR) (Taylor et al., 2007).

15.4.2.5 Esophageal Cancer (EC)

Esophageal cancer (EC) is the eighth leading cancer type worldwide and the sixth leading cause of cancer mortality worldwide. A database review was initiated to investigate the association between meat consumption and esophageal cancer risk. Findings from the present database review reveal that individuals consuming red or processed meat have a 40% increased risk of developing esophageal cancer compared to individuals with low levels of red and processed meat consumption. It was found that each 50 grams of processed meat consumed per day added a 57% increase in the risk of EC (Salehi et al., 2013).

15.4.2.6 Other Cancer

The association between meat and fish intake and the risk of stomach, rectum, pancreas, lung, colon, breast, ovary, kidney, testis, bladder, brain cancer, non-Hodgkin's lymphomas (NHL) and leukemia was investigated. The study took place in 8 Canadian provinces between 1994 and 1997 and included 19,732 cancer cases and 5,039 population controls. It was found that total meat and processed meat consumption were associated with the risk of stomach, rectum, colon, pancreas, lung, breast, bladder, lung, prostate, testis, kidney cancer and leukemia. Specifically, red meat was associated with colon, lung and bladder cancer and processed meat was correlated to stomach, colon, pancreas, rectum, prostate, lung, testis, kidney and bladder cancer and leukemia. It is recommended that the consumption of red and processed meat is limited in order to prevent a number of cancers (Hu et al., 2008).

De Stefani et al., (2012) studied the association between processed meat and the etiology of a number of cancer sites in 6,060 Uruguay participants between 1996 and 2004. Findings reveal a positive association between processed meat consumption and cancers of the rectum, colon, esophagus, lung and stomach. Processed meat is therefore "a powerful multiorgan carcinogen."

15.4.3 Atopic Eczema

The present study investigated the association between maternal fatty acids intake during pregnancy and the risk of atopic eczema among Japanese infants in 771 mother-child pairs. Findings reveal that higher meat consumption during pregnancy is positively correlated with an increased risk in eczema among Japanese infants (Saito et al., 2010).

15.4.4 Stroke

Stroke is a significant cause of mortality and the major cause of disability. The direct and indirect cost of stroke in the U.S. amounted to $73.7 billion in 2010 and $27 billion in Europe in 2010. Study findings revealed that consumption of red and/or processed meat was not only associated with increased risk of ischemic stroke, but also with an increased risk of blood pressure and hypertension, the major risk factors for stroke. Red meat is a significant source of heme iron which increases the risk for atherosclerosis, diabetes and heart disease, all risk factors of stroke (Chen & Wong, 2004). Heme iron can compromise the healthy functioning of the digestive tract leading to DNA damage in cells (Salehi et al., 2013). Processed meat intake exhibited greater risk of stroke compared to red meat consumption. Reducing red and processed meat consumption and increasing healthy food consumption such as fruits, green leafy vegetables and fish may decrease the risk of stroke while also reducing the risk of diabetes, cardiovascular disease and several types of cancers (Chen & Wong, 2004).

15.4.5 Age-Related Macular Degeneration (AMD)

Age-related macular degeneration (AMD) is the number one cause of blindness among older people. Smoking is the only modifiable risk factor of AMD. It is estimated that the number of Americans with AMD will increase by 50% by 2020 to a total of 3 million cases. Similar increases are expected to occur among the Australian population. Study findings revealed that higher red meat consumption was positively association to

AMD. High red meat consumption has been linked with higher levels of N-nitroso compounds, heme iron and glycation end products which could be toxic to the retina due to the oxidative cellular damage induced by them. Oxidative damage is the leading cause of AMD (Chong et al., 2009).

15.5 Meat Consumption

The Italian National Research Institute for Food and Nutrition, the United States Department of Agriculture Center for Nutrition Policy and Promotion, the American Heart Association and the World Cancer Research Funds recommend animal meat consumption (Del Bo et al., 2013). Urbanization, income growth and population growth have contributed to rising meat demands worldwide (Sanda & Oancea, 2012), especially in developing countries stricken by poverty and malnutrition (Webb & Erasmus, 2013). The global average meat consumption per year per person has increased by 12 kg from 1996 until 1999. It is expected that meat consumption will continue increasing, especially in developing countries (Kappeler et al., 2013).

15.5.1 Chinese Population

The Chinese population represents roughly one-fifth of the global population. Improved economic conditions in China since 1978 have resulted in increased meat consumption. Rising meat demand requires the establishment of structural changes in order to expand the livestock industry and therefore increase meat supply to the Chinese population. Expanding livestock industry translates into greater use of land, increased reliance on pesticides and artificial fertilizers and rising agricultural waste, which in turn contributes to climate change, loss of biodiversity, land degradation, air and water pollution. China has developed the fastest-growing food market in the world since its entry into the World Trade

Organization (WTO) in 2001. A study examining meat consumption patterns in China revealed that meat consumption rises as income rises. Increased income also causes the Chinese population to purchase meat with improved quality and safety. Pork meat is the most widely consumed meat in China. Price is the primary obstacle to meat consumption among lower-income Chinese consumers (Liu et al., 2009).

15.5.2 Children

Meat consumption on the day of survey for 4487 children aged 2 to 16 years of age was collected in Australia in 2007. Mean consumption of meat, poultry or fish ranged from 52 g/day in 2-3 year-old boys to 161 g/day in 14-16 year-old boys in Australia in 2007. Mean consumption of meat, poultry or fish in Australian girls between 9 and 16 years of age was 98 g/day. 46% of meat intake was attributed to fish while 38% of energy/day was derived from cereal and cereal-based products (Bowen et al., 2012).

15.5.3 Muslim Goat Meat Consumption

Goat meat demand and consumption is growing drastically in most metropolitan regions in the U.S. such as the Atlanta area due to the increase in immigrants from goat-meat eating parts of the world (Ibrahim et al., 2008).

15.5.4 Consumer Preference

One of the most prominent difference between pasture and concentrated-fed cattle is meat color. Pasture-fed meat has darker color compared to concentrated-fed meat. Meat color is a significant factor in consumer choice and accounts for 15% of the variance of consumer preference for meat. Most consumers prefer bright red meat while a few prefer dark animal meat. A condition known as dark, firm and dry meat (DFD) is prevalent in cattle that have experienced stress and low muscle energy pre-slaughter. 3.5% to 7% of sheep meat contains yellow fat which is not seen as a major concern in New Zealand, yet is perceived as a major problem in Ireland. Meat flavor, derived from over 50 odor-active compounds, is perceived by consumers on the basis of past experiences and cultural background. Meat tenderness is one of the major meat attributes that shapes consumer choice of meat product. Concentrate-fed animals are known to have more tender meat compared to pasture-fed animals. Studies from Ireland reveal that pasture-fed beef with moderate supplementation (i.e., vitamins) produces the most tender meat products (Webb & Erasmus, 2013).

15.6 Meat Labels

The FDA and the USDA require that food labels include accurate nutritional information, amount per serving of saturated fat, cholesterol, dietary fiber and other nutrients of major health implications and nutrient reference values. The attitude of 750 U.S. consumers towards meat labels was investigated. 70% of respondents indicated that food labels assisted them in the purchasing of meat products. 50% of respondents perceived the present information on meat labels as adequate, 30% thought that it was not sufficient. 80% of all respondents believed that it was important that meat labels contain information on meat production process (Rimal, 2005).

15.7 Diary Consumption

There were 25.6 million cows with total milk production of 53 billion kg in 1994 and 9.2 million cows producing 84 billion kg or milk in 2007. Dairy milk production amounts to 2.9% to 4% of total GHG emissions (Webb & Erasmus, 2013).

15.8 Livestock Production

Increasing meat production is perceived as necessary to meet the growing meat demands of a global population that reached 7 billion in 2011 (Sanda & Oancea, 2012). Worldwide animal production is expected to double by 2050 in order to meet food demands. The demand for eggs, milk and meat has been predicted to increase by 30% in the next 8 years (Webb, 2013). The United Nations (UN) estimates that the world population will reach 9 billion people by 2050 which would necessitate the increase of food production being cultivated on the same agricultural lands as is being produced today. FAO states that 70% of the additional food supply needs to be derived from "efficiency-enhancing technologies." Global livestock production enterprises have been criticized for their impact on the environment, animal and human health. Animal agriculture, including dairy production, accounts for about 14.5% of total GHG emissions (Webb & Erasmus, 2013). Agriculture in the U.S. was found to account for 5.8% of GHG emission in the U.S. Livestock agriculture further contributes to loss of environmental biodiversity, environmental degradation and water contamination (Webb, 2013).

Intensification of agricultural production occurs primarily in two directions – horizontal and vertical integration. Horizontal integration incorporates the transformation of natural lands, including conservation and forest areas, into agricultural lands used for livestock grazing. The vertical integration approach, on the other hand, incorporates the increase in the efficiency of animal production. While cattle and sheep in South Africa are primarily cultivated in extensive grazing and pasture systems, cow-calf production systems in Western Europe and North America are often intensive (Webb & Erasmus, 2013).

Animal agriculture in South Africa is characterized as extensive, natural or free range, which is however, prone to

234

weaknesses such as inconsistency in quality and supply, the spread of disease, low productivity and efficiency (Webb, 2013).

15.8.1 Human Rights Violations

A 2005 Human Rights Watch report "condemned" U.S. slaughterhouse and meatpacking companies for significant violations of basic human rights and for "breaching" international agreements regarding safe working environment (Watts, 2005).

15.9 From Animal-Based to Plant-Based Diet

The FAO states that meat and dairy production is to double by 2050 if we are to meet rising demand for animal products. Yet, the environmental impact from livestock agriculture must be reduced into half if we are to prevent escalation of the current ecological damage.

The production and consumption of meat and dairy products pose the greatest environmental risks. A substantial amount of plant material, arable lands, water and raw materials are required to produce meat and dairy products. The transformation of agricultural crops into animal food requires great energy resources (de Bakker & Dagevos, 2012). Livestock nowadays consumes approximately one third of the global production of grain. Further and in light of climate change, a single cow emits 50 kg of methane, a GHG more powerful than CO_2, per year.

Some propose that discontinuing livestock agriculture could contribute to over 15% GHG reduction. Shifting away from livestock agriculture would contribute to additional benefits, including preservation of biodiversity, freeing of land of more than 3 million km^2, reduction of land regions required for agricultural activities, prevention of soil erosion, reduction of nitrogen (N) loading on ecosystems and improved management of phosphorous processes (Allenby, 2012).

Many environmental experts doubt that technological advances in livestock agriculture and more efficient animal production methods will solve the environmental and social issues related to livestock agriculture and global food security. Experts warn about enormous environmental consequences if we fail to convert our diet from animal to plant-based. "The message is clear: a 'technological fix' will not suffice, we also need a 'behavioral' or 'cultural' fix." Technological advances will not be able to meet the growing meat and dairy food products demand. Lowering our intake of meat products while shifting to a more plant-based diet is a solution to such "global issue" (de Bakker & Dagevos, 2012).

Discussion

The adverse health effects from consuming meat are enumerable. One can only imagine the profit acquired by pharmaceutical and health care sectors from health impairments caused by meat consumption and contamination. The grandiosity of profit would explain why animal production companies do not solve meat contamination at its root, but rather devise superficial solutions which only worsen the problem and increase profit. Similarly to air, water, fish, climate and agriculture, meat production is owned by a limited number of large companies supported by governments, increasing industrial and governmental power over humanity and Earthly resources.

Moreover, governments would not be suggesting meat consumption as part of a healthy diet (when in fact it is not healthy) if it was not part of the problem-profit-power scheme. Such a scheme explains why a plant-based diet is not recommended instead, especially given the horrendous conditions under which animals are raised, the availability of plant-based foods, the damage that livestock production exerts on the environment and the adverse health effects resulting from meat consumption.

References

Allenby, B. (2012). A critique of geoengineering. *IEEE Potentials, 12,* 22-26.

Alaejos, M. S., Gonzalez, V. & Afonso, A. M. (2008). Exposure to heterocyclic aromatic amines from the consumption of cooked red meat and its effect on human cancer risk: A review. *Food Additives and Contaminants, 25*(1), 2-24.

Blank, M. M., Wentzensen, N., Murphy, M. A., Hollenbeck, A. & Park, Y. (2012). Dietary fat intake and risk of ovarian cancer in the NIH-AARP Diet and Health Study. *British Journal of Cancer, 106,* 596-602.

Bollaerts, K., Messens, W., Aerts, M., Dewulf, J., Maes, D., Grijspeerdt, K. & Van der Stede, Y. V. (2010). Evaluation of scenarios for reducing human salmonellosis through household consumption of fresh minced pork meat. *Risk Analysis, 30*(5), 853-865.

Bowen, J., Baird, D., Syrette, J., Noakes, M. & Baghurst, K. (2012). Consumption of beef/veal/lamb in Australian children: Intake, nutrient contribution and comparison with other meat, poultry and fish categories. *Nutrition & Dietetics, 69*(2), 1-16.

Bruijnis, M. R. N., Meijboom, F. L. B. & Stassen, E. N. (2013). Longevity as an animal welfare issue applied to the case of foot disorders in dairy cattle. *Journal of Agricultural & Environmental Ethics, 26,* 191-205.

Chen, G.-H. & Wong, M.-T. (2004). Impact of increased chloride concentration on nitrifying-activated sludge cultures. *Journal of Environmental Engineering, 130*(2), 116-125.

Chong, E. W.-T., Simpson, J. A., Robman, L. D., Hodge, A. M., Aung, K. Z., English, D. R., Giles, G. G. & Guymer, R. H. (2009). Red meat and chicken consumption and its association with age-related macular degeneration. *American Journal of Epidemiology, 169*(7), 867-876.

de Bakker, E. & Dagevos, H. (2012). Reducing meat consumption in today's consumer society: Questioning the citizen-consumer gap. *Journal of Agricultural & Environmental Ethics, 25*, 877-894.

Del Bo, C., Simonetti, P., Gardana, C., Riso, P., Lucchini, G. & Ciappellano, S. (2013). Horse meat consumption affects iron status, lipid profile and fatty acid composition of red blood cells in healthy volunteers. *International Journal of Food Sciences and Nutrition, 64*(2),147-154.

De Stefani, E., Boffetta, P., Ronco, A. L., Deneo-Pellegrini, H., Correa, P., Acosta, G., Mendilaharsu, M., Luaces, M. E. & Silva, C. (2012). Processed meat consumption and risk of cancer: A multisite case–control study in Uruguay. *British Journal of Cancer, 107*, 1584-1588.

Eichholzer, M. & Bisig, B. (2000). Daily consumption of (red) meat or meat products in Switzerland: Results of the 1992/93 Swiss Health Survey. *European Journal of Clinical Nutrition, 54*, 136-142.

Heid, A. & Hamm, U. (2012). Consumer attitudes towards alternatives to piglet castration without pain relief in organic farming: Qualitative results from Germany. *Journal of Agricultural & Environmental Ethics, 25*, 687-706.

Hopkins, M. E. & Nunn, C. L. (2007). A global gap analysis of infectious agents in wild primates. *Diversity and Distributions, 13*, 561-572.

Hu, H., Tellez-Rojo, M., Bellinger, D., Smith, D., Ettinger, A. S., Lamadrid-Figueroa, H., Schwartz, J., Schnaas, L. Mercado-Garcia, A. & Hernandez-Avila, M. (2006). Fetal lead exposure at each stage of pregnancy as a predictor of infant mental development. *Environmental Health Perspectives, 114*(11), 1730-1735.

Ibrahim, M., Liu, X. & Nelson, M. (2008). A pilot study of halal goat-meat consumption in Atlanta, Georgia. *Journal of Food Distribution Research, 39*(1), 84-91.

Kappeler, R., Eichholzer, M. & Rohrmann, S. (2013). Meat consumption and diet quality and mortality in NHANES

III. *European Journal of Clinical Nutrition, 67,* 598-606.

Liu, H., Parton, K. A., Zhou, Zhang-Yue & Cox, R. (2009). At-home meat consumption in China: An empirical study. *The Australian Journal of Agricultural and Resource Economics, 53,* 485-501.

Rimal, A. (2005). Meat labels: Consumer attitude and meat consumption pattern. *International Journal of Consumer Studies, 29*(1), 47-54.

Saito, K., Yokoyama, T., Miyake, Y., Sasaki, S., Tanaka, K., Ohya, Y. & Hirota, Y. (2010). Maternal meat and fat consumption during pregnancy and suspected atopic eczema in Japanese infants aged 3–4 months: The Osaka Maternal and Child Health Study. *Pediatric Allergy and Immunology, 21,* 38-46.

Salehi, M., Moradi-Lakeh, M., Salehi, M. H., Nojomi, M. & Kolahdooz, F. (2013). Meat, fish, and esophageal cancer risk: A systematic review and dose-response meta-analysis. *Nutrition Reviews, 71*(5), 257-267.

Sanda, G. M. & Oancea, M. (2012). Analysis of poultry meat consumption in Romania. *Agricultural Management, 14*(2), 171-178.

Sapkota, A. R., Curriero, F. C., Gibson, K. E. & Schwab, K. J. (2007). Antibiotic-resistant enterococci and fecal indicators in surface water and groundwater impacted by a concentrated swine feeding operation. *Environmental Health Perspectives, 115*(7), 1040-1045.

Schellenberg, R. S., Tan, B. J. K., Irvine, J. D., Stockdale, D. R., Gajadhar, A. A., Serhir, B., Botha, J., Armstrong, C. A., Woods, S. A., Blondeau, J. M. & McNab, T. L. (2003). An outbreak of Trichinellosis due to consumption of bear meat infected with *Trichinella nativa,* in 2 Northern Saskatchewan communities. *The Journal of Infectious Diseases, 188,* 835-843.

Stuart, D. & Woroosz, M. R. (2013). The myth of efficiency: Technology and ethics in industrial food production. *Journal of Agricultural & Environmental Ethics, 26,* 231-

256.

Taylor, E. F., Burley, V. J., Greenwood, D. C. & Cade, J. E. (2007). Meat consumption and risk of breast cancer in the UK Women's Cohort Study. *British Journal of Cancer*, *96*, 1139-1146.

Vaarst, M. & Alroe, H. F. (2011). Concepts of animal health and welfare in organic livestock systems. *Journal of Agriculture & Environmental Ethics*, *25*, 333-347.

Vetouli, T., Lund, V. & Kaufmann, B. (2012). Farmers' attitude towards animal welfare aspects and their practice in organic dairy calf rearing: A case study in selected Nordic farms. *Journal of Agricultural & Environmental Ethics*, *25*, 349-364.

Watts, M. J. (2005). Righteous oil? Human rights, the oil complex, and corporate social responsibility. *Annual Review of Environment & Resources*, *30*, 373-407.

Webb, E. C. (2013). The ethics of meat production and quality - a South African perspective. *South African Journal of Animal Science*, *43*(5), S2-S10.

Webb, E. C. & Erasmus, L. J. (2013). The effect of production system and management practices on the quality of meat products from ruminant livestock. *South African Journal of Animal Science*, *43*(3), 413-423.

Zivotofsky, A. Z. (2012). Government regulations of *Shechita* (Jewish religious slaughter) in the twenty-first century: Are they ethical? *Journal of Agricultural & Environmental Ethics*, *25*, 747-763.

Chapter 16

GHOSTLY HABITAT

Overview

"Ghostly Habitat" elaborates on legal hunting. A specific example with leopard hunting is introduced.

16.1 Hunting

Hunting, fishing and agriculture "affect both the size and structure of many animal and plant populations." Exploitation of animal and plant resources "can be used to control population size and expansion" (Gamelon et al., 2012). "Wildlife managers often rely on resource users, such as recreational or commercial hunters, to achieve management goals." Using hunters to control wildlife populations is particularly common, for predator and ungulate populations. Mammalian predators are often harvested as trophies or for fur. The primary motivation, however, is population control. "The recreational or commercial pursuit of a variety of species is used by authorities as a cost-efficient means to control wild populations and thus mitigate direct economic losses" (Bischof et al., 2012).

16.1.1 Leopard

"Sport hunting is often proposed as a tool to support the conservation of large carnivores" (Jorge et al., 2013). Jorge et al., (2013) investigated the economic revenues associated with sport hunting and poaching of leopards (Panthera pardus), the cost of depredation of livestock induced by leopards and people's perception of leopards in Niassa National Reserve (NNR), Mozambique. Findings revealed that leopard hunting was at the cornerstone of the hunting industry. The mean daily hunting safari expenditure for government taxes, licenses, safari expenses, accommodations, trophy management and air charter in NNR accounted to $2,587. The cost of one leopard accounted to U.S. $24,000 while the overall revenue to stakeholders from the legal hunting of a single leopard accounted to U.S. $33,783. 47% of revenue generated from leopard hunting in NNR was retained by the hunting operators, 24% was allocated to the taxidermists and 13% was retained by travel agencies. "Most safari revenues are retained at national and international levels." Only 20% of sport hunting revenues, specifically trophy and concession fees, were legally allocated to villages within the area. The cost of an illegally poached leopard accounted for U.S. $83 or a monthly salary in NNR. People who have had their livestock depredated by leopards endorsed negative attitudes towards leopards. Revenues generated from sport hunting did not compensate for depredated by leopards livestock at the household level. On the basis of such findings, it is proposed that poaching rates are reduced by allocating sport hunting revenues to initiatives that improve household livelihoods, providing villagers with alternative sources of revenue and establishing incentives not to poach leopards.

Conservation of large carnivores is financially demanding and often times competing with alternative societal priorities. Incentives geared towards the conservation of such species may increase the likelihood that conservation efforts will be successful. Substantial revenues from sport hunting are allocated to the conservation of species, habitats and ecosystems in NNR.

16.1.2 Wild Boar & Red Deer

200-270 thousand wild ungulates (i.e., wild boar; red deer) and 400-500 thousand small game species (i.e., brown hare; ring-necked pheasant) have been harvested on a yearly basis in Hungary in the last five years (Bodnar & Bodnar, 2014) while 10,000 tons of produced big game meat from hunting in Spain was derived from red deer and wild boar between 2002 and 2008 (Morales et al., 2011).

16.1.3 Goose

The Svalbard pink-footed goose population in Denmark increased from 33,000 to 50,000 between 1996 and 2005 while the number of harvested geese has expanded from 15,000 in 1997 to 30,000 at present. Evidence suggests that "for certain species nearly one bird was wounded for every one killed." It has been demonstrated that between 28% and 62% of X-rayed goose individuals contained "embedded shot" while this percentage is between 25% and 35% for sea ducks. A national action plan to reduce the number of wounded game species recommended limiting the maximum range for shooting geese in Denmark to 25 meters. The plan has been overall successful in reducing the number of wounded goose populations (Noer et al., 2007).

16.1.4 Wild Animals

"Harvesting wild animals through hunting has become a major conservation issue, especially for large-bodied animals." 114 mammal species out of a total of 350 mammal species are hunted in India. "As a result, there is loss of pollinators, seed predators, seed dispersers and predators, which finally results in an empty forest syndrome." Arunachal Pradesh, the largest state in India, encompasses 2.5% of the country's territory. Harvesting wild animals such as common and clouded leopard, marbled cat, leopard cat, orange-bellied squirrel and yellow-throated marten for subsistence (55%), medicine (10%) and commercial purposes (25%) is common in the area. 33 mammal species are being hunted in Arunachal Pradesh, 57% of which are either endangered, threatened or vulnerable species. The orange-bellied Himalayan squirrel in particular is being harvested for medicine and "social ceremonies" (Selvan et al., 2013).

Further, "the current scale of bushmeat hunting in tropical forests is considered a major threat to biological diversity and forest integrity, and its sustainability is a concern for the livelihoods of millions of the rural poor, who rely on wildlife as a major source of protein and income" (Coad et al., 2013).

16.1.5 Predators

Top predators such as gray wolves play an imperative role in biodiversity. However, predators compete with human beings for resources and space, threatening livelihoods and lives (i.e., domestic animal attacks). The hostility towards predators could contribute to their poaching and objection to conservation efforts which subsequently serve to prevent recovery of predator species (Treves et al., 2013).

16.1.6 Jaguar

The major threats to jaguars (Panthera onca) in northern Mexico are habitat loss and fragmentation, illegal hunting and predator control and depletion of prey species (Rosas-Rosas & Valdez, 2010).

16.1.7 Invasive Animal Species

According to IUCN Species Survival Commission, after habitat loss, invasive species are considered to be the second major threat to native species. Recent evidence demonstrates that deteriorated ecosystems result in "the dominance of alien species over native ones." Game hunting is one of the major causes of the expansion of ungulates species worldwide. High densities of ungulate populations pose a significant threat to plant species due to overgrazing pressures. Habitat degradation and fragmentation in addition to uncontrolled poaching and overexploitation posed a significant threat to native European ungulate populations in the past. "However, current hunting regulations have led to their recovery and even expansion in most countries." Such expansions are particularly evident in areas where sport hunting is not permitted. "Wild ungulates are increasingly being established in regions outside their natural distribution range due to human interests."

The exotic ungulate specie aoudad (Ammotragus lervia) was introduced in south-eastern Spain as game specie in 1970. The native Iberian ibex (Capra pyrenaica) and exotic aoudad have been competing for resources in the south of Europe as aoudad populations have successfully adapted and expanded since their introduction. Despite the fact that current evidences "do not yet" reveal "straight threats," the authors propose the "eradication of the species" as a preventative management measure designed to "preserve the unique natural resources present in this European region" (Acevedo et al., 2007).

Discussion

It could be deduced that there is no convincing evidence to support the theory that animal population control is an actual necessity. The greatest irony is that humanity has been so well deceived that it did not only believe such an illusion, it actually actively participated in the management of animal population control for the government and for the industries, calling it a game – game hunting. The question became as to who is going to win the hunting game – the hunter or the animal, and not whether animal population control is moral or a necessity to begin with. It is apparent how, in line with Sosé's theory, a problem is created (i.e., the need for animal population control) in order for profit to be acquired. The legal hunting of just one leopard generates a profit of U.S. $33,783. The majority of that amount is directed at governmental accounts. The hunting industry, travel and lodging industries are additional sectors profiting from legal hunting.

References

Acevedo, P., Cassinello, J., Hortal, J. & Gortazar, C. (2007). Invasive exotic aoudad (*Ammotragus lervia*) as a major threat to native Iberian ibex (*Capra pyrenaica*): A habitat suitability model approach. *Diversity and Distributions, 13,* 587-597.

Bischof, R., Nilsen, E. B., Broseth, H., Mannil, P., Ozolins, J. & Linnell, J. D. C. (2012). Implementation uncertainty when using recreational hunting to manage carnivores. *Journal of Applied Ecology, 49,* 824-832.

Bodnar, E. S. & Bodnar, K. (2014). Main traits of the wild boar meat in its marketing. *Agricultural Management, 16*(2), 81-86.

Coad, L., Schleicher, J., Milner-Gulland, E. J., Marthews, T. R., Starkey, M., Manica, A., Balmford, A., Mbombe, W., Bineni, T. R. D. & Abernethy, K. A. (2013). Social and ecological change over a decade in a village hunting system, central Gabon. *Conservation Biology, 27*(2), 270-280.

Gamelon, M., Gaillard, J.-M., Servanty, S., Gimenez, O., Toigo, C., Baubet, E., Klein, F. & Lebreton, J.-D. (2012). Making use of harvest information to examine alternative management scenarios: A body weight- structured model for wild boar. *Journal of Applied Ecology, 49,* 833-841.

Jorge, A. A., Vanak, A. T., Thaker, M., Begg, C. & Slotow, A. R. (2013). Costs and benefits of the presence of leopards to the sport-hunting industry and local communities in Niassa National Reserve, Mozambique. *Conservation Biology, 27*(4), 832-843.

Morales, J. S. S., Rojas, R. M., Perez-Rodriguez, F., Casas, A. A. & Lopez, M. A. A. (2011). Risk assessment of the lead intake by consumption of red deer and wild boar meat in Southern Spain. *Food Additives and Contaminants, 28*(8), 1021-1033.

Noer, H., Madsen, J. & Hartmann, P. (2007). Reducing wounding of game by shotgun hunting: Effects of a Danish action plan on pink-footed geese. *Journal of Applied Ecology, 44,* 653-662.

Rosas-Rosas, O. C. & Valdez, R. (2010). The Role of landowners in jaguar conservation in Sonora, Mexico. *Conservation Biology, 24*(2), 366-371.

Selvan, K. M., Veeraswami, G. G., Habib, B. & Lyngdoh, S. (2013). Losing threatened and rare wildlife to hunting in Ziro valley, Arunachal Pradesh, India. *Current Science, 104*(11), 1492-1495.

Treves, A., Naughon-Treves, L. & Shelley, V. (2013). Longitudinal analysis of attitudes toward wolves. *Conservation Biology, 27*(2), 315-323.

Chapter 17

UNFATHOMABLE

Overview

"Unfathomable" paints a picture of elephant poaching and illegal ivory trade. Pertinent policies and regulations are elaborated upon. Elephants' browsing behavior relevant to vegetation diversity is discussed.

***

The deterioration of economic conditions contributes to increased "threats to species verexploitation,...intensification of subsistence poaching and greater reliance on natural resources" (Wittemyer, 2011). Megaherbivores are herbivores such as elephants and rhinoceroses that weigh more than 1000 kg. The "evidence of their ecological significance has been growing. Yet, we are far from understanding the extent to which megafauna extirpation affects biodiversity and ecosystem function" (Sekar & Sukumar, 2013). "The decline in large herbivores across the globe in recent years poses a significant threat to ecosystem integrity" (Kuiper & Parker, 2014). "Large animals are disproportionately likely to go extinct" as a result of the factors contributing to the "ongoing extinction crisis" (Sekar & Sukumar, 2013). Habitat degradation and human population expansion pose challenges to biodiversity conservation efforts. Human and wildlife populations are increasingly competing for land and natural resources (Tingvold et al., 2013).

17.1 Wildlife Trade

The illegal wildlife trade is increasing exponentially around the world, amounting to $20 billion per annum. The largest market for illegal wildlife trade is China followed by the United States (U.S.) and Japan. The Internet is being introduced as an international wildlife illegal trade platform. Illegal wildlife trade is considered "low priority" crime by law enforcement, prosecutor and judicial system officials and is therefore becoming "a high-profit enterprise with exceptionally low risks." Prosecution of illegal wildlife trade traffickers is "rare." Seized large contrabands of snake skins, corals, ivory and conch shells have been described as "the largest of this type in history" (Wasser et al., 2008).

17.2 Elephant Poaching

Between 246,000 and 300,000 elephants reside in South Africa, 118,000-163,000 in Eastern Africa, 16,500-196,000 in Central Africa and 5,500-13,200 in West Africa (Aleper & Moe, 2006). The Convention on International Trade in Endangered Species (CITES) banned the international trade in African elephant ivory in 1989 as a result of the decline of elephant populations from 1.3 million to 600,000 between 1979 and 1987 due to poaching. Poaching rates decreased during the first several years following the ivory ban, yet began increasing thereafter throughout Africa (Comstock et al., 2003).

17.2.1 Civil Conflict

17.2.1.1 Uganda

60,000 elephants lived in Uganda prior to the 1970s. This number in Uganda was reduced by 90% between 1971 and 1980 following the 1971 military coup which resulted in "a breakdown in law and order" (Aleper & Moe, 2006).

17.2.1.2 Angola

"Civil conflict devastates both human and wildlife populations worldwide." Angola's civil war continued for 27 years before it ended in 2002. More than 4 million people were displaced as a result of the war. 100,000 African elephants in Angola were reported killed in the 1980s. The military base of the National Union for the Total Independence of Angola (UNITA) was established in the Luiana Partial Reserve (PR) conservation area. UNITA military as well as people displaced as a result of the civil conflict hunted elephants at Luiana PR and in wildlife areas. Ivory was sold in exchange for ammunition, arms and food. UNITA military was also reported to poach rhinos and sell rhino horns in exchange for arms and food. Several surveys revealed that elephant populations are increasing in density in Luiana PR from 366 in January 2004 to 1827 in November 2005 (Chase & Griffin, 2011).

17.2.2 Elephant Poaching Effects

Poaching contributes to changes in elephants' sex ratio, age structure, social behavior, reproductive patterns and the number of elephants without tusks (Owens & Owens, 2009). Poaching has an impact on the future of elephant populations (Theuerkauf & Ellenberg, 2000). The genetic and social composition of elephants are influenced by human activities. A reduction of genetic variation and a disruption in social relationships result directly from human activities such as poaching (Archie & Chiyo, 2012). Elephants are long-lived. Their tusks continue to grow throughout their lifespan (Pilgram & Western, 1986). Adult elephants are typically preferred for hunting due to their relatively large tusks and greater weight or meat (Wittemyer, 2011). Poachers initially selected adult male elephants followed by adult female elephants for their ivory (Gobush et al., 2008). Male elephant tusks weight substantially more than female elephant tusks (Pilgram & Western, 1986). Increased rates of elephant population declines "can result in random loss of favorable alleles and fixation of deleterious mutations" (Okello et al., 2008).

"Unfortunately, poaching is again on the rise in a number of elephant populations" (Archie & Chiyo, 2012). Several southern African countries have furthermore expressed a "continued interest" in "relaxing the ivory ban" which raises concerns that this would further increase elephant poaching across Africa (Comstock et al., 2003).

17.2.2.1 Ivory: Illegal Trade

Only between August 2005 and August 2006, more than 25,000 kg of ivory have been seized which is 3 times the total volume seized in the past 3 years. Ivory is considered as a "general goods" item. Other items belonging to the "general good" compartment are bootlegged CDs and DVDs. Seizure rate for "general goods" contraband is 10% as opposed to a much higher rate for drugs and explosives. Therefore, the 25,000 kg of ivory seized between August 2005 and August 2006 correspond to 250,000 kg of smuggled ivory. 250,000 kg of ivory implies the poaching of 38,000 elephants. 38,000 elephants represent 8% of the remaining 470,000 elephants in Africa. This is a higher percentage than the 7.4% offtake of elephants per year which took place 10 years prior to the ivory ban in 1989. At an 8% annual poaching rate, elephants in sub-Saharan Africa will be extinct by 2020. The price of ivory increased from $200/kg in 2004 to $850/kg in 2006. A solution to the illegal ivory trade crisis is to direct enforcement officials towards the areas where wildlife is being poached, "stopping poaching at its source." Forensic investigators have been largely unable to locate the geographic origin of illegal ivory as their point of reference begins at the point of shipping as opposed to the point of its source. It is not necessary that poached ivory is exported from the same geographical location from which elephants are poached

(Wasser et al., 2008). The geographical origin of illegal elephant ivory is established by extracting DNA from ivory (Comstock et al., 2003).

Conservation authorities have been mainly concerned regarding the illegal and unsustainable offtake rates of African elephants in the past 2 decades. Two attempts have been undertaken to assess the impact of the elephant poaching ban on elephant poaching since 1990. The lack of reliable data on the allocation of law-enforcement resources and related illegal poaching rates resulted in statistically inconclusive findings (Jachmann & Billiouw, 1997).

17.2.2.1.1 Regulation

CITES is responsible for measuring the rate and trends of illegal hunting of elephants (Kahindi et al., 2009). The majority of countries producing ivory are signatories of the CITES which permits the export of elephant products provided "'such export will not be detrimental to the survival of that species'" (Pilgram & Western, 1986).

17.2.3 Human-Elephant Conflict

Hunting elephants for ivory and meat and transforming habitats to agricultural land are the primary causes for the "local extinction" of the African elephant "over much of its former range" (Kuiper & Parker, 2014). A number of anthropogenic factors affect animals outside of protected areas where they must co-exist with human populations, livestock and agriculture. Illegal killing and human-elephant conflict are the primary anthropogenic factors affecting elephant populations (Ahlering et al., 2013). The International Union for Conservation of Nature (IUCN) defines human-elephant conflict as: "Any human-elephant interaction which results in negative effects on human social, economic, or cultural life, on elephant conservation or on the environment"

(Tingvold et al., 2013). Human-elephant conflict increases as human populations settle throughout elephant habitats. Commercial forestry, mining and petroleum operations require the building of roads in previously inaccessible habitat areas hence contributing to the increase in elephant poaching rates (Walsh et al., 2001). Human-wildlife conflict and habitat fragmentation is furthermore constraining elephants "to live in smaller areas with growing contact with humans and livestock" (Archie & Chiyo, 2012).

17.2.3.1 Luiana PR

Re-colonization of elephant populations in Luiana PR and the general future of wildlife in the south-east parts of Angola may be compromised by several factors. First, the Caprivi Border Fence, "a double electrified veterinary fence" erected along the Botswana's northern border with Namibia in 1997, in addition to human settlement in proximity to the border prevents the movement of wildlife. Secondly, human populations displaced by the civil war are returning, settling and establishing agricultural practices in Luiana PR, posing an additional restrain to elephants' movement while also enhancing human-elephant conflict. Human settlement in Luiana PR conservation area is further projected to prevent effective poaching control. Third, unexploded landmines from UNITA military centers remain in Luiana PR, posing additional threat to elephant populations. 30 elephants were reported killed by landmines in southern Angola. Lastly, Angola is "emerging as an important source of illegal ivory" (Chase & Griffin, 2011).

17.2.4 Conservation Areas

The continuous decline of elephant populations has contributed to the establishment of enclosed, usually fenced areas, designed to afford protection to elephants (Kuiper & Parker, 2014). Conservation areas are typically too small to provide long-term protection and adequate space for animal species that require large habitat areas. Human-elephant conflict transpires alongside the boundaries of protected areas. Property damage, depredation of livestock, prohibited access to previously available natural resources and the risk of being killed contribute to the human-elephant conflict (Tingvold et al., 2013). More than 80% of elephant habitats are located outside of conservation areas due to elephant population numbers expanding beyond the boundaries of protected areas (Ahlering et al., 2013).

17.2.4.1 Biological Diversity

"The conservation of biological diversity is one of the greatest challenges facing humanity today." Preventing extinction of species, intervening to preserve the structure, composition, function, heterogeneity of existing species, restoring degraded ecosystems, are all strategies designed to contribute to the conservation of biological diversity. Dense populations of large herbivores may compromise the biological diversity of habitats (Kuiper & Parker, 2014). "To compensate for the energy requirements of their enormous bodies, elephants need a great deal of living space and abundant vegetation and water" (Demeke et al., 2012). Elephants enclosed in confined areas have been found to compromise the biological diversity of habitat plant species (Kuiper & Parker, 2014). The diet of elephants has been found to consist of as many as 146 plant species. Elephants consume greater than any herbivore amount of vegetation (Staub et al., 2013). The spatial distribution of African elephants is regulated by the availability of water points (Fullman & Child, 2012). Enclosing vegetation and changing water locations are two methods proposed to limit the damage on vegetation (Staub et al., 2013).

According to Kuiper & Parker (2014), on the other hand, "change within ecosystems is natural and inevitable. Large herbivores have been living in an ever-changing environment for millions of years." Habitat changes incurred by elephants "should not automatically be categorized as undesirable and short term changes may be part of an overarching trend of long-term stability." The vegetation damages incurred by elephants may therefore represent "a shift towards a more natural historical state." The scientific view that elephants are "biodiversity thieves...appears to be based primarily on intuition, rather than on hard science." Much of the habitat degradation incurred by elephants is the direct result of building artificial water centers and fences to constrain elephants in confined areas. Such "unnatural" confinement "to localised areas" have "disrupted seasonal movement,...disturbed the natural dynamics of population processes" and therefore contributed to the localised impact of elephants on vegetation structure and composition. Tree fall caused by elephants was found to be six times greater in enclosed as opposed to excluded areas in Kruger National Park in South Africa. Elephants in Tanzania's Ruaha National Park were shown to contribute to the death of 37% of Acacia albidia and 67% of Commiphora ugogenesis trees.

17.2.5 Seed Dispersal

Megaherbivores are especially important seed dispersers as they are able to swallow large and small vegetation seeds such as fruit seeds without impairing them (Sekar & Sukumar, 2013). Forest elephants (Loxodonta africana cyclotis) are important vegetation seed dispersers, thereafter contributing to the regeneration of vegetation (Theuerkauf & Ellenberg, 2000). At least 13 plant species in African forests are particularly adapted for dispersal by elephants. For example, the fruits of *dillenia indica*, also referred to as "elephant wood apple," are consumed by elephants, yet are unlikely to be accessed by smaller animal species. Free-ranging elephants are also particularly effective seed dispersals due to the extensive land areas over which they are likely to deposit vegetation seeds (Sekar & Sukumar, 2013).

17.2.6 Stress

Stress is defined as "an environmental stimuli that leads to an imbalance of homeostasis as the 'stressor,' and the corresponding defense reaction of an animal as the 'stress response'". Wildlife animals experience chronic stress when they are unable to adapt their behavior to anthropogenic risk factors. Chronic stress contributes to lower reproduction and survival rates (Tingvold et al., 2013).

African elephants change their behavior as a result of their psychological state. Elephants engage in reclusive behavior (i.e., restricted exploratory movement; long-distance allocation) and aggression towards human beings in response to increased stress hormone levels (Jachowski et al., 2013). Hunting risk increases the stress hormone levels in elephants (Tingvold et al., 2013). "A disrupted family group is a chronic stress condition for African elephants." Chronic stress and chronically elevated glucocorticoids inhibit elephant immunity, growth and reproduction (Gobush et al., 2008).

Estimating the faecal glucocorticoid metabolites (FGM) is a noninvasive method generally used to monitor fluctuations in adrenocortical function as a result of stress. The faecal glucocorticoid metabolite levels of African elephants residing in areas with low and high human interference were compared. Results revealed that elephants had higher glucocorticoid levels in areas with greater as opposed to lower degrees of human interference. Particularly, lower levels of glucocorticoid were found in elephants residing within a protected area. Still, a great number of human beings residing alongside the boundaries of the protected areas admitted that they have been engaged in hunting activities within the protected area (Tingvold et al., 2013).

Discussion

"Unfathomable" proposes that economic scarcity (i.e., problem) contributes to the over-exploitation of species and natural resources. The reality of economic scarcity is not surprising given the highly concentrated nature of profit and power (i.e., water privatization; aquaculture; agriculture, etc.). As "Downward Acceleration" will also suggest, global wealth is increasingly being concentrated in the hands of a few. Declining economic conditions therefore predispose for the destruction of the environment, which, similarly to past examples, results in increased profit opportunities for the few.

Furthermore, illegal ivory is perceived by authorities as a general goods item. Transporting illegal ivory is equated to transporting bootleged CDs and DVDs. Drug smuggling is regarded as a much greater crime compared to the smuggling of illegal ivory. How humanity chooses to treat their own bodies is regarded as a crime of a high degree while elephant mutilation is compared to the illegal burning of a CD or a DVD. Though research on the topic of the war on drugs has not been included as part of "Sovereign Terra," previously established patterns and examples of how problems are created to generate profit and power allow us to deduce that there is much greater profit and power generated from the war on drugs compared to the war on illegal ivory which explains why the war on drugs is a priority.

References

Ahlering, M. A., Maldonado, J. E., Eggert, L. S., Fleischer, R. C., Western, D. & Brown, J. L. (2013). Conservation outside protected areas and the effect of human-dominated landscapes on stress hormones in savannah elephants. *Conservation Biology, 27*(3), 569-575.

Aleper, D. & Moe, S. R. (2006). The African savannah elephant population in Kidepo Valley National Park, Uganda: Changes in size and structure from 1967 to 2000. *African*

Journal of Ecology, 44, 157-164.

Archie, E. A. & Chiyo, P. I. (2012). Elephant behaviour and conservation: Social relationships, the effects of poaching, and genetic tools for management. *Molecular Ecology, 21*, 765-778.

Chase, M. J. & Griffin, C. R. (2011). Elephants of south-east Angola in war and peace: Their decline, re-colonization and recent status. *African Journal of Ecology, 49*, 353-361.

Comstock, K. E., Ostrander, E. A. & Wasser, S. K. (2003). Amplifying nuclear and mitochondrial DNA from African elephant ivory: A tool for monitoring the ivory trade. *Conservation Biology, 17*(6), 1840-1843.

Demeke, Y., Renfree, M. B. & Short, R. V. (2012). Historical range and movements of the elephants in Babile Elephant Sanctuary, Ethiopia. *African Journal of Ecology*, 50, 439-445.

Fullman, T. J. & Child, B. (2012). Water distribution at local and landscape scales affects tree utilization by elephants in Chobe National Park, Botswana. *African Journal of Ecology, 51*, 235-243.

Gobush, K. S., Mutayoba, B. M. & Wasser, S. K. (2008). Long-term impacts of poaching on relatedness, stress physiology, and reproductive output of adult female African elephants. *Conservation Biology, 22*(6), 1590-1599.

Jachmann, H. & Billiouw, M. (1997), Elephant poaching and law enforcement in the central Luangwa Valley, Zambia. *Journal of Applied Ecology, 34*, 233-244.

Jachowski, D. S., Montgomery, R. A., Slotow, R. & Millspaugh, J. J. (2013). Unravelling complex associations between physiological state and movement of African elephants. *Functional Ecology, 27*, 1166-1175.

Kahindi, O., Wittemyer, G., King, J., Ihwagi, F., Omondi, P. & Douglas-Hamilton, I. (2009). Employing participatory surveys to monitor the illegal killing of elephants across diverse land uses in Laikipia-Samburu, Kenya. *African*

Journal of Ecology, 48, 972-983.

Kuiper, T. R. & Parker, D. M. (2014). Elephants in Africa: Big, grey biodiversity thieves? *South African Journal of Science, 110*(3/4), 1-3.

Okello, J. B. A., Wittemyer, G., Rasmussen, H. B., Arctander, P., Nyakaana, S., Douglas-Hamilton, I. & Siegismund, H. R. (2008). Effective population size dynamics reveal impacts of historic climatic events and recent anthropogenic pressure in African elephants. *Molecular Ecology, 17,* 3788-3799.

Owens, M. J. & Owens, D. (2009). Early age reproduction in female savanna elephants (Loxodonta africana) after severe poaching. *African Journal of Ecology, 47,* 214-222.

Pilgram, T. & Western, D. (1986). Inferring hunting patterns on African elephants from tusks in the international ivory trade. *Journal of Applied Ecology, 23,* 503-514.

Sekar, N. & Sukumar, R. (2013). Waiting for Gajah: An elephant mutualist's contingency plan for an endangered megafaunal disperser. *Journal of Ecology, 101,* 1379-1388.

Staub, C. G., Binford, M. W. & Stevens, F. R. (2013). Elephant herbivory in Majete Wildlife Reserve, Malawi. *African Journal of Ecology, 51,* 536-543.

Theuerkauf, J. & Ellenberg, H. (2000). Movements and defaecation of forest elephants in the moist semi-deciduous Bossematie¨ Forest Reserve, Ivory Coast. *African Journal of Ecology, 38,* 258-261.

Tingvold, H. G., Fyumagwa, R., Bech, C., Baardsen, L. F., Rosenlund, H. & Roskaft, E. (2013). Determining adrenocortical activity as a measure of stress in African elephants (*Loxodonta africana*) in relation to human activities in Serengeti ecosystem. *African Journal of Ecology, 51,* 580-589.

Walsh, P. D., White, L. J. T., Mbina, C., Idiata, D., Mihindou, Y., Maisels, F. & Thibault, M. (2001). Estimates of forest elephant abundance: Projecting the relationship between precision and effort. *Journal of Applied Ecology, 38,* 217-

228.

Wasser, S. K., Clark, W. J., Drori, O., Kisamo, E. S., Mailand, C., Mutayoba, B. & Stehens, M. (2008). Combating the illegal trade in African elephant ivory with DNA forensics. *Conservation Biology, 22*(4), 1065-1071.

Wittemyer, G. (2011). Effects of economic downturns on mortality of wild African elephants. *Conservation Biology, 25*(5), 1002-1009.

Chapter 18

NANO MANIPULATION

Overview

"Nano Manipulation" discusses nano technology, including recent developments, health hazards and regulations. The chapter concludes with an elaboration of nanobiosensor technology.

***

"The food industry is ultimately driven by profitability, which is consequent on gaining consumer acceptance". Improving food taste, freshness, shelf-life, traceability and nutritional value, reducing cost, manufacturing foods geared towards specific groups of people or diets, ensuring sustainability and safety of food production are all means to increase consumer acceptance.

"The food industry is already suspected in some quarters of secretly using nanotechnology in their products." It is therefore imperative for industries to involve the public in decision-making processes regarding the use of nanotechnology, including the issue of labeling food containing nanoparticles. Nanotechnology is a multibillion dollar industry, led by the United States (U.S.), followed by Japan and China and expected to had generated 1 trillion dollars by 2015. The global nanotechnology market employs 2 million workers. 400 companies are presently applying nanotechnologies to food. This number is expected to continue growing exponentially.

Nanotechnology is designed to improve food taste, texture, nutrient absorption and packaging. Patent databases contain "more than 450 patent entries with regard to applications of nanotechnology in food." As many as 150 nanotechnology products are under development. Nanoencapsulated substances, activated under certain conditions such as under particular microwave frequencies, are also under development. Several nanotechnology products, including nutritional supplements and additives are commercially available and expected to grow exponentially worldwide.

Food packaging materials obtained from nanotechnology comprise the largest application of nanotechnologies in the food sector. Miller Brewing Co. (USA) and Hite brewery Co. (South Korea) have been reported to use nanoclay, a nanotechnology derivative, in their beer bottles. Nanosilver, a different nanotechnology substance, known for its antimicrobial properties, is used in food storage containers, baby milk bottles, plastic storage bags and refrigerators to preclude microbial growth. A possible risk from the application of nanotechnology in food packaging is the possibility of migration of nanoparticles into food or beverages. Data on migration of nanoparticles to food and beverages is lacking, yet food packaging products derived from nanotechnology are commercially available at present (Chaudhry et al., 2008).

18.1. Human and Environmental Health Effects

Data on the effects of nanoparticles on the environment and more specifically the bioaccumulation and concentration of nanoparticles in the food chain is further insufficient. Data on toxicology and consumer health pertinent to nano-based food and food packaging is generally lacking. Major concerns result from the lack of knowledge on the interaction of nanoparticles at the molecular level and the resulting on human health and the environment effects. The rapid increase of nanotechnologies in a number of consumer products in addition to lack of knowledge relative to nanotechnologies has raised environmental, safety, ethical, regulatory and policy concerns. Further, the dividing line between food, cosmetics and medicine is unclear and further obscured by the ability of nanoparticles to interact with biological organisms "at a near-molecular level."

Nanoparticles may endanger human health upon entry into the stomach via the consumption of food and beverages containing nanoparticles. Data on the effects of nanoparticles entering the gut via the gastrointestinal route is scarce. Scientific findings reveal that nanoparticles can cross cellular barriers and lead to potential oxidative damage to cells in addition to "impairment of DNA replication and transcription." Particles as small as 70 nm can enter the cell nucleus. Nano-sized food additives and substances can further cross the stomach wall, leading to additional likelihood for health consequences. A different concern pertains to the potential of nano-based dietary supplements contained in nanoparticles to enter the blood and introduce foreign to the blood substances.

A common denominator between the three main components of food, carbohydrates, proteins and lipids, is that digestion of the three constituents of food occurs at a nano-scale level. "The nano-scale processing of foods may alter how the food ingredients 'behave' upon breakdown within the gut and, as a consequence, how they are treated in the GI tract." The intestinal mucus, responsible for confining and removing foreign materials before they come in contact with the gut epithelium, surrounds the epithelial cells. Particle size and surface charge determines passage and rate of passage through the mucus layer. Smaller particles are more easily diffused through the mucus layer compared to larger particles. As soon as food constituents are broken down by enzymes, diffused through the intestinal mucus layer and absorbed by epithelial cells, they are transported across cells and directed to the hepatic portal circulation. Possible link has been established between micro and nanoparticals and the worsening of symptoms in people with impaired GI function. It is unknown as to whether micro or nanoparticles elicit a similar inflammatory response in people with healthy GI. Studies investigating as to whether a reduction of microparticles intake from diet can reduce inflammation responses in people with impaired GI function have generated contradictory results (Chaudhry et al., 2008).

18.2 Regulation

Additives are substances added to food in order to achieve a particular purpose such as food preservation. Additives in the European Union (EU) are authorized for use by the European Parliament and Council. The European Legislation has established a pre-market approval process for novel foods, defined as foods with a new molecular structure or foods with significantly altered structure, nutritional values, metabolism or level of undesirable substances. Foods containing nanoparticles would not fall under the "novel" food category as they are not substantially structurally different from their historical or traditional food counterparts. A requirement of the EU Food Law Regulation 178/2002 is that food released in the market is safe for consumption. The vendor offering the food or food packaging for sale is ultimately responsible for ensuring food or food packaging safety. The application of the precautionary principle relevant to nanotechnology has been proposed considering the lack of legislation specifically dealing with nanoparticles, the lack of sufficient toxicology data on nanoparticles and the possible harm caused by nanoparticles (Chaudhry et al., 2008).

18.3 Nanobiosensor Technology

New technology in the form of a chip "will enable to provide a collection of a frankly staggering amount of individual data." Nanobiosensor technology which allows for in vivo and real time monitoring and diagnosis, could be combined with the use of chips. Nanotechnology and microfluidics serve to create nanobiosensor technology. "The entire sensor could be downsized to the nanoscale." The European Technology Platform on NanoMedicine expects the commercial release of "multifunctional nanoparticles for drug release and imaging" in 2011 and "implantable devices for continuous measurement of blood markers" around 2015. Nanobiosensor in vivo monitoring "will be used to inform individuals about their health status, including propensity to disease/medical conditions long before they occur." "One expects soon implantable bio-sensors which will deliver round the clock, instant and seemingly complete information on one's entire physical being". Nanobiosensor technology is "intended to become...part of routine health check practice" used on a global scale for "'common' disease prevention." The technology is intended to "be used to predict, prevent, treat, or ease disease development." Efficacy and cost reduction are two factors proposed as "very strong incentives to promote the mass distribution of the nanobiosensor technology" (De Tavernier, 2012).

18.3.1 Paternalism

Paternalism, originating in the second half of the eighteenth century, is defined in the Oxford English Dictionary as "the principle and practice of paternal administration; government as by a father; the claim or attempt to supply the needs or to regulate the life of a nation or community in the same way a father does those of his children." Paternalism is defined by the authors as "the intentional overriding of the consumer's known preferences or actions, where the institution who overrides justifies the action by the goal of benefiting or avoiding harm to the consumer whose preferences or actions are overridden." Paternalism therefore allows governments to limit and override free or autonomous choices "even if it involves force to intervene in another person's preferences and actions." The question that arises is: "Under what circumstances and to what extent may we give up autonomy and respect for autonomous choices in order to benefit the beneficence of individual persons" (De Tavernier, 2012)?

Discussion

"Nano Manipulation" proposes that nano technologies are being secretly used in products while labeling products containing nano particles is not required. Secrecy and lack of proper labeling (i.e., deceit) would not be necessary unless nano particles are poisonous to the human being ingesting them. Nano particles have the capacity to cross the cell membrane and perform a certain function within the cell itself. They have built-in intelligence. One nano particle is coded to improve the taste of food, another - texture and so on and so forth. It becomes apparent that the range of possibilities of coding nano particles are vast. Secrecy ensures purchase of nano products which generates profit for the multi-billion dollar nano industry. The poisonous nature of nano particles creates additional opportunities for profit for the pharmaceutical and health care sectors.

Similarly to GMO technology, governmental regulations pertaining to nano technology are based upon the principle of substantial equivalence which ensures continual profit for nano technology industries. Profit for governments and industries is further guaranteed when the vendor and not the producer of foods and food packaging materials containing nano particles is held responsible for food and food packaging product safety (i.e., transferring responsibility). Similarly to GM crops which were commercially released despite the fact that research is inconclusive, secretive and highly controversial, nano particles are being patented and commercially released without adequate research proving safety.

Comparable conclusions could be drawn for nanobiosensor technology. If nano particles could be programmed to perform a wide variety of functions pertaining to food and food packaging, then the possibility that nanobiosensor chips could be programmed to perform additional functions to the ones consumers are made aware of is also present. Deriving from Sosé's problem-profit theory, such functions could include emotional, cognitive, behavioral monitoring, cellular and genetic manipulation, to name a few. The greater the acquired information pertaining to an individual, the greater the opportunity for creating problems tied to profit and power over the individual.

References

Chaudhry, Q., Scotter, M., Blackburn, J., Ross, B., Boxall, A., Caste, L., Aitken, R. & Watkins, R. (2008). Applications and implications of nanotechnologies for the food sector. *Food Additives and Contaminants*, *25*(3), 241-258.

De Tavernier, J. (2012). Food citizenship: Is there a duty for responsible consumption? *Journal of Agricultural & Environmental Ethics*, *25*, 895-907.

Chapter 19

"NON-PROFIT" SECTOR & FOOD "AID"

Overview

The structure, objectives and pertinent to non-profit organizations issues and trajectories are presented. Food insecurity and hunger are elaborated on. History, policies, regulations, goals and problems arising from the food aid sector are discussed. Solutions to established by the food-aid sector problems are noted.

19.1 Non-Profit Organizations

19.1.1 Non-Profit Principal & Agent

A recent study investigated the goals of nonprofit managers (principals) and employees (agents) and the possible challenges of the nonprofit principal-agent relationship. Findings reveal that the principal and the agent often have different objectives which may prevent the agent to "act in the principal's interests." Adverse selection and moral hazard are two challenges characterizing the nonprofit principal-agent relationship. Such problems occur "when the behavior of the agent is not controlled or restrained." Adverse selection corresponds to the "misrepresentation of skills and abilities by the agent" resulting from the inability of the principal to "observe the relevant characteristics of the potential agent before signing the contract". Moral hazard, on the other hand, occurs when the agent may "misbehave" and fail to "put forth the agreed-upon effort." Several authors have concluded that "the lack of efficient principals and difficulties with output measurement" contribute to such nonprofit agency challenges. Principals are perceived as less motivated to monitor the performance of nonprofit agents as the "exercise of control over nonprofit organizations is not linked to claims on profits." "Second, given the complex objectives and hard-to-observe outputs of most nonprofit organizations, nonprofit boards may have difficulties finding effective performance criteria to serve as basis for calculating managerial remuneration" (Van Puyvelde et al., 2013).

19.1.2 Non-Profit Director & Chair

Board chairs and executive directors are responsible for the development of the nonprofit organizations to which they belong. Executive directors bear responsibility for the daily operational activities of their nonprofit organizations while board chairs are primarily responsible for long-term policy and policy clarification. Yet, "the distribution of roles and resources between these actors is empirically and normatively highly disputed." The power relationship between board chairs and boards of directors is a "critical aspect of governance." Power in such context is defined as "the ability to execute one's will against resistance." Increasing organizational and environmental pressure on board chairs and executive directors (i.e., on governance) "affects the power relation and the likelihood that one of the actors will be put into a position to carry out his or her own will despite resistance from the other" (Jager & Rehli, 2012).

19.1.3 Governmental & Non-Profit Sector

In general terms, the public or governmental sector is characterized by "multiple layers of bureaucracy," comprehensive rules regarding pay and promotion and enhanced management structure while nonprofit organizations have less emphasis on management structure, fewer promotion opportunities and "less transparent systems determining pay and promotion." Nonprofit organizations are further "subject to less external political influence and public scrutiny" compared to public or government agencies. Managers in public organizations "often work in a culture of conflicting values" due to the "diverse demands made by various actors in the political environment." Compared to managers employed by the public sectors, managers in nonprofit organizations report increased degree of freedom in terms of deciding how to perform their job duties, greater control over their work schedule and more opportunities for pay increase. The results of surveys reveal that the majority of recent college graduates prefer working for nonprofit organizations than for the government sector (LeRoux & Feeney, 2013).

19.1.4 Non-Profit & Profit Sector

"Nonprofits represent the fastest-growing sector in the U.S. economy." Income generated from commercial activities is the "fastest-growing source of nonprofit revenues" (LeRoux & Feeney, 2013). Nonprofit literature increasingly suggests that "the lines dividing for-profit and nonprofit organizations are becoming increasingly blurred...Nonprofit organizations now operate in an environment in which they compete for resources and are required to prove as well as to improve their effectiveness." More sophisticated strategic and control managerial practices are thereby increasingly being put in place to "improve efficiency and productivity" (Tucker & Parker, 2013). More and more managers in nonprofit organizations are holding business degrees (LeRoux & Feeney, 2013).

One study investigated for-profit management practices within 32 Australian nonprofit organizations. Findings of the study revealed that managerial control and strategy in nonprofit organizations is similar to that of profit organizations, yet for different reasons. Two implications emerge from the findings of the investigation. First, "coercive pressures on nonprofit organizations" may contribute to the operation of parallel control systems within nonprofit organizations and thereby the "duplication of scarce resources." Secondly, the nonprofit application of a corporate, "business mind-set" model that focuses on managerial control and strategy "immediately clashes with the philanthropic values" of a number of nonprofit organizations (Tucker & Parker, 2013).

19.1.5 Non-Profit Wages

"Wages influence motivation, work satisfaction, productivity, worker fluctuations, cost, and profit." Study findings reveal that the recruitment of volunteers in Austrian nonprofit organizations contributes to employees working for nonprofit organizations receiving lower wages (Pennerstorfer & Trukeschitz, 2012).

19.2 Food Insecurity

"Peace and the welfare of human society depend fundamentally on a sufficient, balanced, and secure supply of food" (Schmidt & Wei, 2006). There is a common consensus that the current rates of global food insecurity are "unacceptable morally and ethically as well as from a social and economic development policy perspective." "The current state of world food insecurity is a dark stain on the record of international cooperation given that the tools and technologies to mitigate food insecurity among the most vulnerable are well proven, widely available, and inexpensive." It is imperative that policymakers seek innovative ways to establish consensus among diverging norms and rules that have previously been a source of conflict that impeded efforts towards strengthening the worldwide governance of food security (Margulis, 2013).

840 million people were estimated to be food insecure

in 2013 (Lentz & Barrett, 2014). It has been estimated that 10.1% of United States (U.S.) households or 31 million Americans are food insecure. From the 31 million food insecure Americans, 3 million households experience hunger (Wehler et al., 2004).

Food security does not pertain solely to food availability and the ability to produce food. Problems related to access to food are "critically tied" to food ownership and exchange, which are "exacerbated" in the contexts of economic crisis and unpredictability, inequality and poverty.

Food crisis was prevalent in a number of countries worldwide even before food prices began increasing globally and exponentially since 2008. It is projected that purchasing power for food and other essentials will continue to be low or fall even lower as a result of the "unfolding global financial crisis and economic slowdown." Developing food-importing, low-income, food-aid depending countries such as Haiti are most susceptible to food price shocks (Conceicao & Mendoza, 2009).

Global food prices are predicted to steadily increase as climate change continues to have a detrimental impact on the environment. Prices of processed foods are expected to increase by 20% by the year 2030. The poorest and marginally food-insecure households are most vulnerable to food price increases. Food price increases are expected to cause food-secure households to become marginally food-insecure. Access to "nutritious, safe, affordable, and enjoyable food is a prerequisite for health." Any lack thereof could contribute to the development of chronic disease. 78% of health care expenditures in the U.S. are allocated to the treatment of chronic disease (Dodge, 2013). A link has been established between high-calorie processed food consumption and obesity risk (Husband, 2013). Obesity, the second leading cause of illness and death after tobacco use in the U.S., is correlated to marginally food-insecure households. It contributes to the development of type 2 diabetes and cardiovascular disease which are both chronic conditions (Dodge, 2013).

19.2.1 Obesity

Obesity results from greater food energy intake compared to energy use. It is a global, ever growing, health-care taxing endemic which affects developed as well as developing countries, adults as well as children, likely compromising worker productivity (Alston et al., 2006). Obesity "has become one of the most serious but neglected global public health problems, especially in developing countries" (Asfaw, 2006). Over 1.5 billion people in the world are overweight. Obesity rates have doubled worldwide since the 1980s. Obesity and "associated" diabetes rates are exponentially growing worldwide (Basu et al., 2013).

19.2.1.1 The U.S.

The U.S. has one of the highest rates of obesity in the world. Obesity in the U.S. and the resulting health impairments have become priority concerns for the U.S. government. $113 billion were spent by the U.S. Department of Agriculture (USDA) on food stamps, educational, environmental, subsidized school lunch and domestic agriculture programs in 2004 as a means to reduce obesity. While lower agricultural commodities prices have been established to reduce hunger, they have also contributed to an increase in the consumption of foods high in fat and/or sugar which are prerequisite to the development of obesity. One proposed solution to reducing obesity rates in the U.S. is the implementation of fiscal and regulation tools that decrease "unhealthy" food (i.e., foods high in fat or sugar) choices by increasing their prices through added taxation while decreasing the consumer costs of "healthy" foods (i.e., fruits and vegetables) and increasing their quality (Alston et al., 2006).

19.2.1.2 Egypt

Alarmed by increasing prevalence of diabetes, the WHO issued a report in 2004, calling for research on the contributing to obesity factors, particularly in developing countries. Yet, little research has been undertaken to explore the root causes of obesity in the developing world.

     70% of women and 48% of men were either overweight or obese in Egypt in 1998. Obesity rate in Egypt, particularly amongst women, is one of the highest in the world. Based on data from 1997, the present study explored the relationship between a government food subsidy program and obesity rates in mothers in Egypt. The food subsidy program which costs the Egyptian government US $1.1 billion in 1997 makes bread, wheat flour, oil and sugar available at subsidized prices. Results reveal that obesity amongst mothers is generally not dependent on income. "Income increases the level of body mass index (BMI) but at a decreasing rate." 1% increase in the price of bread was likely to decrease BMI in mothers by 0.12% while 1% increase in the price of sugar was likely to decrease mothers' BMI by 0.16%. On the other hand, 1% decrease in the price of fruit was likely to reduce mothers' BMI by 0.08% and 1% decrease in the price of milk and eggs was found to likely reduce BMI of mothers by 0.12%. The government food subsidy program provides foods rich in carbohydrates, starch and fat at reduced prices, yet does not influence prices of micronutritient-rich foods. This contributes to high intake of energy-dense foods and therefore increased rates of obesity. Even though the Egyptian government has undertaken an initiative to educate the Egyptian population on healthy diet, obesity and chronic conditions resulting from obesity, the authors believe that additional strategies such as providing micronutrient-rich food products at reduced price and shifting the focus of the program from "ensuring basic food items" to "ensuring basic healthy food items" should be considered in order to reduce Egyptian mother's BMI (Asfaw, 2006).

19.2.1.3 Soft Drink Consumption

Previous studies reveal that soft drink consumption in the U.S. has been correlated to increasing rates of obesity and diabetes as soft drinks contain large quantities of refined sugars which "contribute to excessive weight gain, the metabolic syndrome, and insulin resistance."

The relationship between soft drink consumption and global rates of obesity and diabetes between 1997 and 2010 was explored. Data on over 70 countries were acquired from the Euromonitor Global Market Information Database, the International Diabetes Federation and the WHO. Findings reveal that soft drink consumption increased from 9.5 gallons per capita per year in 1997 to 11.4 gallons per person per year in 2010 worldwide. 54% of soft drink consumption between 1997 and 2010 occurred in developing and middle-income countries. The highest consumption of 37.8 gallons per person of soft drink consumption was recorded in the U.S. in 1998. The soft-drink consumption in the U.S. decreased to 31.2 gallons per capita in 2010. A 1% increase in soft drink consumption worldwide was associated with an additional 2.3 obese adults per 100, 4.8 overweight adults per 100 and 0.3 adults diagnosed with diabetes per 100. Increased soft drink consumption was further associated to improvements in income. In sum, soft drink consumption was found to be "significantly" associated to diabetes, obesity and overweight worldwide, including developing and middle-income countries.

Soft drink consumption is projected to increase by 15.7% in the next 5 years in low and middle-income countries and 9.5% worldwide. Such an increase is equivalent to an additional 2.3 billion adults to be overweight, 192 million new cases of diabetes and 1.1 billion people to become obese (Basu et al., 2013).

19.2.2 Diabetes

Type 2 diabetes mellitus (T2DM) is exponentially increasing on a global level (Rylander et al., 2014). At least 1 in 20 adults globally have diabetes (Basu et al., 2013). 153 million people suffered from diabetes in 1980. The prevalence of diabetes has doubled between 1980 and 2008 (Rylander et al., 2014) and increased from 5.5% in 2000 to 7.0% in 2010 (Basu et al., 2013). Diabetes is projected to increase to 550 million by 2030. The risk factors of diabetes include obesity, overweight, family history of diabetes, compromised glucose tolerance, age, hypertension, sedentary lifestyle and history of gestational diabetes. Cardiovascular disease (CVD) and T2DM share a large number of the same risk factors (Rylander et al., 2014).

The American Diabetes Association endorses a healthy diet rich in fruits, vegetables, grains, low-fat milk and fish for people suffering from diabetes. However, people who suffer from diabetes, especially patients with low income, cannot afford the recommended diet as prices of healthy foods are higher compared to prices of unhealthy (i.e., high in sugar, fat, calories) foods. Food prices are positively correlated to blood glucose levels in low-income people suffering from T2DM and inversely associated to blood glucose levels in high-income diabetics. Health care professionals propose subsidizing healthy foods while taxing unhealthy food products as a means to improve the accessibility of people suffering from T2DM to a healthy diet (Anekwe & Rahkovsky, 2014).

19.3 Poverty

Poverty, natural disasters, population growth, inequality, governance, conflict and social exclusion are some of the causative to food insecurity in the world factors (Perez-Exposito & Klein, 2009). "The long-term drivers of food security are fundamentally intertwined with the roots of poverty and inequality" (Conceicao & Mendoza, 2009).

Poverty alleviation rate increased between 1970 and 1990 and significantly declined since 1990 with international aid reaching an "all-time low." Half of the world's poor population is concentrated in rural areas. 75% of poor human beings are considered to be "extremely poor." Natural cataclysmic events, economic shocks and war can rapidly change the number of poor people worldwide. Poor people who own assets (i.e., small farmers with shelter) can quickly become poor human beings with no assets (i.e., refugees) (Naughton-Treves et al., 2005). 36.6 million Americans were poor in 1997 while the income of 14.6 million Americans was below the federal poverty level (Kasper et al., 2000). The poverty rate in Bolivia accounts to 42% while the poverty rate in Sierra Leone is 75%. The purchasing power parity (PPP) of people in Nepal, India and Sri Lanka was less than $1.4 per day in 2005. More than 70% of said daily living expenditure is allocated to food. While the priority of the majority of poorest nations in the world relates to food access, developed countries prioritize access to energy (Conceicao & Mendoza, 2009).

19.4 Hunger

Hunger in developing countries does not result from food unavailability. "The world already produces sufficient food." There is sufficient grain produced worldwide to provide each person on Earth with 3,500 calories per day.

Political upheavals in a number of countries in the past 2 decades have contributed to starvation and food deprivation. For example, the war between the Arab populations of North Sudan and South Sudan over oil resources contributed to the death of 2 million Sudanese from famine during a 17-year period. In the 1990s, thousands of Afghans experiencing hunger relocated to Herat, situated near the border of Iran, due to civil war, drought and inadequate government. Hunger is a continual phenomenon in developing countries despite the

Green Revolution marking the introduction of GM crops. Other reasons include discriminatory economic policies against agriculture, inability of farmers to acquire credit and access marketing services, dysfunctional rural infrastructure and restricted market opportunities for farm production. World hunger would not be eliminated provided such problems are not adequately addressed. Increased crop production would be an asset provided such problems are solved. Yet, genetic engineering is not necessary to achieve such increase. Sufficient food can be produced using current agricultural methods (Jordan, 2002).

The Executive Director of the Women's Foreign Policy Group stated in 2008: "[w]ithout food, societies become breeding grounds for instability, civil unrest, terrorism and demagogues." Hunger provides the basis for dictators, terrorists and warlords to "manipulate or control the hungry and vulnerable and to influence regional stability and international interests" (Essex, 2012). "The worst violators are those who use food as a weapon, as in Sub-Saharan African conflicts" (Marchione & Messer, 2010).

19.4.1 Prevalence

A decline in the number of hungry people between 1970 and 1995 was followed by a steady growth of the prevalence of hungry people worldwide. Such reversed global trend was not expected and has "surprised most policy-makers" (Margulis, 2013). It has been estimated that only 10% of hungry people in developing countries suffer from acute hunger resulting from natural disasters, wars or "sheer destitution" while the remaining 90% experience chronic hunger leading to malnutrition. The number of chronically or acutely hungry people in developing countries has increased by 109 million since 2004 for a total of 963 million people worldwide (Sanchez, 2009). 923 million people in the world were chronically hungry in 2007 – an increase of 75 million hungry people between 2003 and 2005 (Essex, 2012). The prevalence of hungry people worldwide "swelled" to 1 billion in 2008 as a result of increased food and energy prices, food trade bans, biofuel policies and "speculation on commodities markets." Despite the fact that the number of hungry people has slightly decreased since 2008, food prices sharply increased once again in 2010 and 2011, creating uncertainty in regards to world food supply. "The recent food price crises exacerbated an already deteriorating world food security situation" (Margulis, 2013).

Hunger prevalence is measured according to two main indicators: the undernourished or those populations whose food intake is below the minimum level of energy consumption and the number of underweight children under the age of 5 years (Marchione & Messer, 2010).

19.4.2 Undernourishment

The reason for undernourishment and food insecurity is rather the lack of financial resources to purchase food and the inaccessibility to agricultural resources, technology, education, infrastructure and credit to produce food. Poverty and not the unavailability of food is also the reason people living in urban and rural ghetto areas in developed countries to experience undernourishment.

According to data available from the World Health Organization (WHO), one third of children in the world are malnourished (Jordan, 2002). Over 800 million individuals in developing countries do not have adequate access to food and "the number has increased since 1995" (Hoddinott et al., 2008). "In the wake of soaring food prices, the number of undernourished people increased up to a record high of one billion in 2009." Over 2 billion people experienced micronutrient deficiency prior to the food crisis of 2007/8 (Kuhlgatz & Abdulai, 2012). Mineral and vitamin deficient populations are "more widespread with some 4 to 5 billion people consuming iron-deficient diets." Mineral and vitamin deficiencies can contribute to significant public health consequences (Hoddinott et al., 2008). Children of young age are particularly vulnerable to the effects of malnutrition (Porter, 2010). Malnutrition leads to compromised immune system, the development of diseases and high mortality rates in children (Sanchez, 2009). The first 5 years of life are very important. Adverse events that take place during this period may have "permanent effects." Inadequate nourishment in early childhood, for example, prevents healthy cognitive development which leads to poor performance at school and impaired health in adulthood (Porter, 2010). About 140 million children under the age of five globally are Vitamin A deficient. There are also concerns regarding folic acid and zinc deficiencies among children under the age of five worldwide (Hoddinott et al., 2008). Malnutrition and hunger are primary causes of mortality and morbidity in developing countries, especially among vulnerable populations such as mothers and children. The fundamental nutrients necessary for health include energy, minerals, vitamins, fat, proteins from animal and plant sources. Insufficient consumption of any of these nutrients contributes to adverse health consequences (Kuhlgatz & Abdulai, 2012).

19.4.3 Underweight

More than half of children's mortality from infectious diseases is associated with being moderately to severely underweight. "Underweight children live on the margins of disease, death, and a lifetime of low productivity and chronic disease." The majority of underweight children and undernourished people live in Sub-Saharan Africa, the Pacific and Asia regions (Marchione & Messer, 2010).

19.4.4 Contributing Factors

Contributing to hunger factors include energy and food price changes, financial crisis, "obsolete development policies" (Sanchez, 2009), food supply, availability, use and access (Margulis, 2013). Gender, race, employment status and education are risk factors associated with poverty as well as hunger. It has been demonstrated that mothers' mental and physical health is an additional predictive of hunger factor (Wehler et al., 2004).

19.4.5 Prevention & Intervention

Food access and not solely food supply is critical in preventing famine (Margulis, 2013). More often, hunger results from the inaccessibility to food resources rather food scarcity. Poverty and inequality are the underlying factors that hinder the ability of consumers to produce or access adequate qualities and quantities of food resources (Marchione & Messer, 2010). The major policies associated with hunger prioritize a quick-fix approach to alleviating hunger, failing to take into consideration the causes of hunger (Sanchez, 2009).

The Millennium Development Goal (MDG) was established in September 2000. Its overall objective was to eliminate hunger and extreme poverty. More specifically, it aimed to reduce the number of people who experience hunger by half between 1990 and 2015 (Marchione & Messer, 2010). As set forth by the MDG, the goal of reducing the number of hungry people worldwide by half between 1990 and 2015 is projected to be unmet. World poverty rates remained stable and even declined at the same time that world hunger rates rose. "Rising hunger has occurred alongside constant growth in world food production, in both absolute and per capita terms" (Margulis, 2013). "Technically, there is more than enough food in the world to feed everyone a basic, nutritionally adequate diet" (Marchione & Messer, 2010).

19.5 Fortified Blended Foods (FBFs)

Fortified blended foods (FBFs) containing high micronutrient and protein concentrations were developed in 1967 (Perez-Exposito & Klein, 2009) as part of the Food for Peace Program also known as US Public Law 480 (Fleige et al., 2010) to "improve the nutritional status of children suffering from malnutrition." Initially intended to improve the nutritional needs of moderately malnourished preschool children in the U.S., FBFs have spread internationally as well. The U.S. is the main donor of FBFs in the world. FBFs are distributed via supplementary feeding programs (Perez-Exposito & Klein, 2009). FBFs represent the "most nutrient-dense" Food for Peace Program cereal product (Fleige et al., 2010).

Fortifiers, formulations and nutrient content differ among various FBF products. Corn-soy blend (CSB) has become the most popular form of FBF used since the 1980s. Yet, "no systematic approach has evolved to develop products suited to the nutritional needs of vulnerable populations. FBFs have come to be used as a one-size-fits-all products for a wide array of vulnerable groups," including children, pregnant women, populations living with HIV/AIDS and those exclusively dependent on food aid (Perez-Exposito & Klein, 2009).

19.5.1 FBF Research

"In almost 40 years, very little research has been conducted on them." Although weight gain, "recovery from moderate acute malnutrition" and "prevention of severe micronutrient deficiencies" have been observed among populations receiving FBF food-aid, "direct measurements of the micronutrient status of vulnerable populations receiving FBFs have rarely been conducted. Well-designed evaluations to test the efficacy and effectiveness of FBFs on linear growth and development are, likewise, lacking." "Technical difficulties" in data collection in addition to the variety of FBF programs have prevented rigorous efforts in evaluating the impact of FBF foods on improving nutritional outcomes. "Evidence of the efficacy of fortified blended foods for improving nutritional outcomes is currently limited and weak."

A review of available supplementary feeding programs reports was conducted in 1982. Results revealed that "the programs had no consistent impact on improving the nutritional status of young children" in developing countries (Perez-Exposito & Klein, 2009).

A Food Aid Quality Enhancement Project was designed to evaluate the effectiveness of FBFs in meeting the nutritional needs of vulnerable populations. Findings reveal that "FBFs do not meet the nutritional needs of infants and young children between the ages of 6 and 24 months." Results also indicate that improvements relative to FBFs intended for school-aged children and adults are also necessary. FBFs "one-size-fits-all" concept for all population groups "is not adequate." The development of two FBF variations – one for vulnerable infants and children between 6 and 24 months and one specifically designed to meet the nutritional needs of older children and adults has been recommended.

A 1998 assessment of FBF products revealed "significant variation in the vitamin content at points of manufacture." Despite new requirements in response to the 1998 assessment, "the uniformity of fortification had not sufficiently improved." Further, studies demonstrate a "significant loss of vitamins during cooking of fortified food aid products." Instances of FBFs turning green color when cooked were also demonstrated. Such FBFs quality concerns caused Congress to task the Government Accountability Office (GAO) to assess the Food for Peace Program. "The GAO found quality problems on several overseas site visits" (Fleige et al., 2010).

19.5.2 FBF Modification & Improvement

Few nutritional content changes to FBFs have been made since 1998 (Perez-Exposito & Klein, 2009). Problems such as spoilage of previously used food aid products necessitated the development of an improved food aid product called "instant corn-soy blend." The improved product is manufactured by food technologist Charles Onwulata in the Dairy Processing and Products Research Unit at the ARS Eastern Regional Research Center (ERRC) in Wyndmoor, Pennsylvania and is "particularly for young children." It consists of toasted soy flour, pre-gelatinized cornmeal, deodorized, refined and bleached soybean oil, pre-mixed minerals, vitamins and antioxidants. Each 100-gram serving of the instant corn-soy blend contains a uniform amount of minerals and vitamins, has a neutral to nutty flavor and yellow to golden buff color (Bliss, 2011).

19.6 Food Aid Governance & Logistics

A number of developing countries do not have formal social protections systems put in place. Such nations rely on ad hoc and external support (i.e., food aid donors) during times of crisis (Porter, 2010).

19.6.1 U.S. Food Aid

The establishment of U.S. Public Law (PL) 480 in 1954 initiated the beginning of food aid. There are three types of food aid assistance under PL 480 – Title I, II and III. "Title I is government-to-government aid in the form of concessional sales with the express aim of opening new markets for U.S. grain." Title II assistance represent food aid distributed by nongovernmental organizations (NGOs) and the World Food Programme (WFP) during states of emergency (Clapp, 2005). Title II is "mandated to address famine, combat malnutrition, address causes of hunger, protect the environment, promote economic and community development, carry out feeding programs, promote democracy, and most recently, prevent conflict" (Marchione & Messer, 2010). One of the key objectives of Title II is to "combat malnutrition, especially in children and mothers" (Fleige et al., 2010). "Title III is government-to-government grants of food aid for development activities" (Clapp, 2005). The U.S. federal government spends considerable amount on food assistance each year. Food assistance expenditures accounted to a total of $41.8 billion in 2003 (Kabbani & Kmeid, 2005).

The U.S. is the largest donor of food aid to developing countries (Fleige et al., 2010). The U.S. donates approximately 50% of all international food aid (Kneteman, 2009). U.S.-based international food aid programs between 1946 and 2004 were valued at $73 billion (Marchione & Messer, 2010). It has been estimated that 1 million metric ton of grain would adequately feed 5 million starving people for one year (Svedberg, 1978). U.S. food aid donations ranged between 2.5 and 11 million metric tons between 1989 and 2006. The variation was explained by food prices rather than the worldwide need for food. In 2006, Public Law 480 Title II donated 3.7 million metric tons of food to 50 nations, providing food resources to millions whose lives were threatened as a result of economic, social or climatic shocks (Marchione & Messer, 2010). The U.S. "committed" $2.3 billion to the Food for Peace Program which were distributed in the form of 2.3 million metric tons of food to about 50 million food-insecure individuals in 49 countries primarily under Title II, the largest US Public Law 480 program (Fleige et al., 2010).

"Because of its agricultural abundance and the humanitarian impulses of the American people, the U.S. is the most generous food aid donor country" (Marchione & Messer, 2010). Since the inception of food aid, the U.S. has remained as the major donor country (Clapp, 2005).

"Tied" food aid is defined by the Organisation for Economic Co-Operation and Development (OECD) as "any aid that requires the procurement of goods and/or services from the donor country." Tied food aid can either be in-kind or it can be in the form of "concessional loans contingent on buying food from the donor." In-kind food aid is defined as "raw or processed agricultural products which are donated to a recipient government or non-governmental organisation (NGO)" (Kneteman, 2009).

As opposed to the majority of donor nations donating cash equivalent food aid, the U.S. donates food aid primarily in kind and has thereby a "considerable influence on the content of aid shipments" (Kuhlgatz & Abdulai, 2012). In kind U.S. food aid has consisted of domestically grown wheat, soy and corn (Clapp, 2005). Section 204 of US Public Law 480 requires that 75% of non-emergency agricultural commodities consist of "fortified, processed, and bagged products" (Fleige et al., 2010). Cash equivalents are intended for local purchase of food at recipient countries. The EU shifted their food aid donations from kind to cash-based in the 1990s. Such shift was perceived as promoting the development of international countries (Clapp, 2005).

19.6.2 International Food Aid

The Food Aid Convention (FAC), signed in 1967, is one aspect of international food aid governance (Hoddinott et al., 2008). All major donor countries are members of the FAC. Argentina, Australia, Japan, Canada, Norway, the European Community and its member states, Norway, Switzerland and the U.S. are FAC members (Clapp, 2005). FAC is the only treaty that legally requires its member donor nations to "provide international development assistance" (Hoddinott et al., 2008). FAC determines the minimum amount of food aid, denominated in tons of wheat, that donor countries are to give annually. Food aid can either be given in kind or in cash equivalent. Additional commodities other than wheat are also allowed to be given, yet they are still measured in wheat equivalents (Clapp, 2005).

There are no consequences in instances when donor countries fail to meet their commitments or the agreement of members to supply developing countries with food aid or the cash equivalents. The U.S. food aid was 170,000 metric tons short from its pledge of 4.48 million tons in 1994. Canada has "missed its commitments in four of seven years" since 1999 while Argentina has never met its food aid pledge. FAC does not make the failure of signatory donors to meet commitments public. A different limitation of FAC is the lack of an evaluation system to ensure the effectiveness of food aid or the overall performance of signatories. Further, only FAC signatories are part of the Food Aid Committee. NGOs and aid-recipient countries are excluded from FAC terms negotiations and Food Aid Committee discussions regarding policy and practice.

FAC "operates with little transparency. The Food Aid Committee provides remarkably little public information on its deliberations, even though members are democratic governments that are accountable to their citizens." A different criticism relative to transparency revolves around the expanded set of institutions and actors that take part in food aid governance of which FAC and the Food Aid Committee are part. "Authority is diffuse and mandates overlap, with the result that accountability is often unclear." The London-based International Grains Council (IGC) has been the host agency and secretariat of FAC since the establishment of FAC in 1967. One of the functions of IGC is to collect signatory reports on food aid shipments. "Its data are not publicly available; only signatory governments receive them."

The majority of food aid is presently donated in cash equivalent (Hoddinott et al., 2008). More and more countries including Canada and European countries are providing food aid donations in voucher and cash equivalents which allow aid-recipient countries to rely on local and regional procurement (LRP) "whereby food aid commodities are acquired in recipient or neighboring countries rather than being shipped from the donor country." 13% of all food aid in 1994/5 was

LRP while this number increased to 67% by 2010. LRP, cash and vouchers are considered to be more time and cost effective. A study revealed that on average, LRP, cash or vouchers reduced food aid delivery time by 13.8 weeks compared to transoceanic food aid. The U.S. has remained as one of the very few donor countries providing food aid in kind rather than in cash. U.S. food aid in kind was responsible for 89% of all donated food aid in kind in 2011 (Lentz & Barrett, 2014). Indifferently of the type of food aid, "one of the primary objectives of food aid is poverty alleviation" (Kirwan & McMillan, 2007).

19.6.3 Political Means

"Unfortunately, bilateral food aid donor allocations are not always motivated by altruism and concern for alleviating malnutrition and poverty in recipient countries" (Awokuse, 2011). Title II and other U.S. food aid programs do not perceive human rights as the main motivational factor for donating food (Marchione & Messer, 2010). "On the surface," from the moment of its inception, PL 480 appeared as a humanitarian endeavor designed to provide U.S. food to countries in need of food assistance. Under the surface, however, PL 480 "was clearly a mechanism for surplus disposal and export promotion in the United States." PL 480 served to establish a market for surplus U.S. food which contributed to higher domestic prices of U.S. grain. The U.S. surplus of grain decreased in the mid-1990s, "making the surplus disposal element of food aid appear to be less significant than in the past."

U.S. food aid also functioned as a political tool (Clapp, 2005). "Food aid has always been used to satisfy U.S. geopolitical and security interests" (Marchione & Messer, 2010). PL 480 was amended in the 1960s "to explicitly tie donation of food aid to political goals, in particular to favor noncommunist countries" (Clapp, 2005). During the Cold War, for example, the U.S. used food aid as a political means to "attract and reward anti-Communist allies, no matter what their human rights records" (Marchione & Messer, 2010). Europe and Canada also demonstrated economic and political reasons for donating food aid (i.e., surplus disposal), yet such tendencies were perceived as less overt compared to the U.S. The overt political aims of PL 480 have been removed since the 1980s (Clapp, 2005).

Recent research findings relative to food aid allocations demonstrate that the top food aid recipient nations are not necessarily the countries in most need of food aid (Awokuse, 2011). One study demonstrated that the world's three biggest food aid donors in addition to NGOs exhibit donor interest bias "in the form of preferential treatment of geographically close countries." It was also found that, in reference to both emergency and total aid, nations in receipt of "higher share of US military aid receive more food aid from NGOs" (Neumayer, 2005). Data further reveals that the governments of food aid recipient countries tend to use received food aid as "'payment' for political support" rather than to distribute it among the most malnourished households (Awokuse, 2011).

The effectiveness of the Bosnia universal food aid program between 1994 and 1997 was investigated. Findings revealed that 61% of the distributed food aid did not reach intended households. "Some of the food was diverted to entities other than households, such as the nationalist armies." Food aid can become "a weapon." Additional obstacles to delivering food aid to intended beneficiaries included inadequate vehicles, fuel, staffing, administration, "restricted access to records" and inaccurate information on demographics (Andersson et al., 2012).

Countries that host a larger number of refugees are not as likely to receive total food aid while countries "with voting similarity in the UN general assembly are more likely to receive total food aid." Poorer nations are more likely to receive total and emergency food aid. Countries with greater number of Catholics and Protestants as well as nations which are in geographical proximity to the U.S. or Western Europe are more likely to be eligible for total food aid. "Countries which receive a higher share of US military aid receive more food aid from NGOs...These results hold both for total and emergency aid". The results from this analysis reveal that food aid allocation is not quite free from donor interest. The majority of donor countries prefer to supply food aid to nations in geographical proximity to the donor country, to the U.S. or to Western Europe in instances of WFP and NGO food aid. The geographical proximity bias should not be strictly perceived as an attempt of donor countries to "maintain a regional sphere of influence." Based on study findings, WFP food aid allocation in the 1990s appears to be "quite sensitive" to the food aid need of recipient countries (Neumayer, 2005).

19.6.3.1 China

Between 1960 and 1962, the People's Republic of China "experienced one of the worst famines in world history...Misguided government policies, exacerbated in many areas of the country by problematic weather conditions" contributed to famine and the death of millions of Chinese people. Nationwide famine further caused "deteriorating morale and social problems." The Chinese Communist Party directly contributed to the nationwide starvation by re-directing agricultural efforts to industrial production, requiring "men and women...to leave the fields to work in factories or government-sponsored projects." After a debate within the administration of John F. Kennedy, the final decision of the U.S. was to "ignore pleas" for the provision of food assistance to China. China's economy began improving in 1962 when the economic focus shifted from industrialization to agricultural production. U.S. President John F. Kennedy concluded that "China presented the long term danger to the peace" (Grasso, 2003).

19.6.3.2 Haiti

A 7.0 magnitude earthquake hit close to Haiti's capital Port-au-Prince on January 12, 2010. More than 285,000 meals "packaged two weeks before the earthquake by over 2,000 volunteers in El Dorado, Kansas, was parachuted to locations throughout Haiti by the U.S. Army" and distributed to affected families two days after the earthquake. Meals were packaged by volunteers of the humanitarian food aid organization Numana Inc. Children in Haiti were "at risk" prior to the earthquake and "the uncertainty and instability caused by the aftermath has only exacerbated this situation." 1 in 2 children in Haiti were in school before the earthquake and over 1.2 million children were "deemed to be extremely vulnerable to violence, exploitation and abuse." "To make clear that Numana meals should not be exchanged for children or money, labels were attached to the bags that indicated in Creole and English that the meals were free and not to be sold; labels also warranted against child trafficking" (Ballard-Reisch, 2011).

19.6.3.3 U.S. Food Aid Domestic Market

The U.S. purchases its food aid "mainly from large agribusinesses" on the domestic market. The US Department of Agriculture's Farm Service Agency (FSA) is responsible for food aid procurement. "FSA's strict procurement regulations permit only a small number of pre-qualified, U.S.-based agribusinesses to bid for government food aid contracts." FSA's regulations "increase profits for the select group of eligible firms." Compared to open-market prices, such regulations cause the U.S. government to pay 70% more for corn and on average 11% more for all food aid commodities. More than half of all U.S. food aid was purchased from four U.S. agricultural industries and five U.S. shipping companies accounted for more than one half of all U.S. food aid freight expenditures in 2005. Except for the U.S., all industrialized World Food Programme donor countries have either fully or partially untied their food aid. The United States Department of Agriculture (USDA), the United States Agency for International Development (USAID) and "surprisingly" President George W. Bush have proposed at least partial untying of U.S. food aid. Such suggestions have been however met with substantial resistance from the "iron triangle" of food aid: domestic food processor, shipping and farm groups as well as from NGOs which depend on monetizing food aid to fund development projects. Such "lobbying pressure" has contributed to U.S. Congress and Senate to "flatly" refuse "any proposals to untie aid." Studies reveal that tied food aid reduces the effectiveness of food aid by approximately 35%. Tied food aid "has been criticised as an implicit form of export subsidy that governments use to circumvent export subsidy restrictions" (Kneteman, 2009).

19.6.4 Food Aid Dependency

"Food aid is unreliable and has not delivered long-term developmental benefits to the poorest countries" (Kirwan & McMillan, 2007). Food aid is commonly viewed as a "necessary evil:" necessary to alleviate hunger in food-deficit households, yet evil as it undermines domestic agricultural production incentives, contributing instead to reliance on food aid. Almost all studies agree that food aid has a depressing effect on domestic food prices and that such consequence is "almost always harmful to producers" (Tadesse & Shively, 2009). Governments of aid-receiving nations may be less inclined to invest "scarce resources" on agricultural development when they are aware that they can depend on food aid during crisis times. The "dumping" of surplus food either for free or at a low cost to developing countries contributes to farmers in aid-receiving nations losing the ability to produce food at competitive prices or losing the incentive for agricultural production. This eventually leads to the "deterioration of the infrastructure of production" in aid-receiving countries (Kirwan & McMillan, 2007).

In 2006, the U.S. donated $1.2 billion in food aid to Africa, only $60 million of which were devoted to agricultural development in Africa (Sanchez, 2009).

19.6.5 Food Aid Decline

Several developments "impinge on future food aid." For example, growing amounts of grains produced in North America "are likely to go into producing alternative fuels." Approximately 20% of U.S. maize production in 2008 is projected to be converted into ethanol fuel "instead of being available for world food markets or as a food aid resource." Next, controversies regarding GM food aid are likely to continue. Third, climate change is expected to contribute to declines in agricultural production in developing countries and to more extreme and frequent weather events such as floods and droughts, creating the need for food aid. Lastly, declining support for domestic agriculture may contribute to reduced availability of in-kind food aid (Hoddinott et al., 2008).

"Food aid programs have been an isolated and diminishing resource for addressing hunger." By the 21st century, worldwide food aid is "reaching no more than one-fifth of hungry people each year" (Marchione & Messer, 2010). 68% of the UN humanitarian aid appeals were covered by donors in 2005 while only 53% of the humanitarian assistance appeals for people impacted by violence in Sudan were covered by donor countries. "In per capita terms, food aid to all developing countries has fallen by two-thirds since 1970" (Hoddinott et al., 2008).

19.6.6 Emergency Food Aid

Food aid falls into two main categories: programs that assist developing countries in becoming sustainable and programs that provide immediate assistance during periods of food scarcity (Toft, 2012). Food emergency resources have grown since the 1980s while food aid has transformed into "a resource of last resort" (Marchione & Messer, 2010). The U.S. food aid average budget of $2.2 billion in the past decade is not sufficient to provide assistance to all food insecure individuals. Consequently, the U.S. government (USG), similar to other donor countries, has increasingly focused its assistance on populations affected by conflict or natural disasters (Lentz & Barrett, 2014). Food for Peace donates $2 billion worth of commodity-based food aid annually during emergencies (Bliss, 2011). Twice as much emergency food aid is directed towards conflict-related emergencies when compared to natural disaster-related emergencies (Hoddinott et al., 2008). "Images of children are frequently used to garner support for emergency food-aid programmes in rich donor countries." Yet, little is known about the impact of emergency food aid on children's health (Porter, 2010).

19.6.7 Food Aid Ineffective

19.6.7.1 Governance

The governance of international food aid consists of institutions and agreements that have been described as "dysfunctional and outdated" and therefore "ineffective" (Hoddinott et al., 2008). The period between shipment of U.S. food aid and delivery to recipient countries is 5 months. Such length of time "may have negative effects on the timeliness proper targeting of food aid when it is most needed" (Awokuse, 2011).

19.6.7.2 Monetization

Monetization of U.S. food aid is the process by which U.S. food aid institutions sell U.S. food aid in aid-recipient countries to raise cash necessary for food security projects. More than one half of Title II non-emergency food aid was accounted for by monetization practices in the past two decades. Monetization has been criticized for wasting millions of U.S. taxpayer dollars and for disrupting regional markets. Title II food aid monetization had 75% recovery rate in FY 2012 which was equivalent to a waste of $32 million of taxpayer dollars. $32 million dollars are sufficient to provide food assistance to more than 800,000 people. Most donor countries have therefore removed monetization practices from food aid policies. "Greater transparency about grossly inefficient monetization events could curtail them and may help make the case for more cash-based assistance."

50% of all Title II food aid is shipped on American-flagged vessels. Shipping U.S. food aid on U.S.-flagged vessels in FY 2006 costs 46% more than shipping the food donations at competitive freight prices. In FY 2012, U.S. taxpayers spent more Food for Peace aid dollars on transport and handling (45%) than on actual food (40%). The findings from a study revealed that "politicians who received more than $10,000 from shipping groups voted against reforming food aid by 7 to 1, noting 'money talks.'" Approximately 10,000 food-deficient people are not being assisted for every U.S. shipping job protected. As opposed to Title II food aid, Canada manages to spend approximately 70% of its food aid budget on commodities including LRP, cash and vouchers (Lentz & Barrett, 2014).

19.6.7.3 Program Effect

Quantitative as well as qualitative research methods are essential in evaluating the effectiveness of food aid programs. While there are some qualitative studies, empirical studies are generally lacking. The scarcity of quantitative data substantially constraints a comprehensive evaluation of the effectiveness of food aid programs (Awokuse, 2011). A number of studies have investigated the association between food assistance and food security, yet have only slightly succeeded in establishing a significant program effect (Kabbani & Kmeid, 2005).

International agreements including the Food Aid Convention measures food aid efforts in grain equivalents rather in nutrient sufficiency. Food aid consists primarily of cereals indicative by "an average 86 per cent share of cereal products in global food aid for the years 1993 to 2007." While being energy-rich, cereal products are being criticized for failing to meet the nutritional requirements of vulnerable populations. "The human metabolism needs nutrients, and not certain product types." A primary objective of food aid is to "reduce malnutrition and attain specific nutritional targets at the operational level."

The effectiveness of food aid in meeting the nutritional requirements of aid-recipient populations was investigated. Energy, iron, zinc and vitamin A are "severely deficient in developing countries" and were therefore the focus of exploration. The study examined the food aid supplies from the U.S. and non-U.S. donor nations between 1993 and 2007. Study findings revealed that food aid has been primarily geared towards nutrient and budget-deficit populations. More specifically, non-U.S. food aid donors responded more sensitively to the nutritional needs of regions affected by crisis while "monetization practices in US project food aid seem to give NGOs incentives to focus on development projects primarily in rural and politically stable regions." "The US food aid does not adequately respond to fluctuating needs" which calls for untying U.S. food aid. Findings from the study further revealed that "donor-related issues significantly affect nutrient allocation...Donor interest bias can be found primarily in US programme and project food aid." Lastly, "cereal data analyses are unable to capture nutrient aid responses" (Kuhlgatz & Abdulai, 2012).

A different study investigated the impact of food aid on

the food supply, nutritional status and dietary behavior of 1664 French food aid users in 2004/5. Findings from the study revealed that food aid was the only source of food for more than 70% of the participants. The mean food aid budget per month accounted to 70.00 Euros per month per capita which is lower than the critical minimal threshold. 46% of the participants were "food-insufficient." Only 1.2% of the participants met the French dietary recommendations for fruits and vegetables, 9.2% - for dairy products, 27.3% - for seafood, 48.7% - for starchy foods and 49.4% - for meat, fish and eggs. 16.7% of the participants were classified as obese, 29.4% had high blood pressure, 14.8% suffered from anemia, 67.9% were at risk for folate deficiency and 85.6% had vitamin D deficiency. The study provides evidence for an unhealthy dietary patterns and poor health among French food aid recipients and points to "the necessity of improving the nutritional quality of currently distributed food aid." "Food aid associations, whose status is private but which depend on food donations from public and private sectors, are themselves prone to strong economic and practical constraints in terms of supplies, distribution and storage" (Castetbon et al., 2011).

19.6.7.4 Tied Food Aid

Tied food aid is "less efficient than untied aid and depressed local agricultural production in recipient countries...56% of the labour force in developing countries is engaged in agriculture." Tied food aid that discourages local food production is also projected to prevent local farmers from consumption (Kneteman, 2009).

19.7 Solution

A recent study reveals that improvements in targeting food aid is the most significant factor in determining the effectiveness of food aid (Awokuse, 2011). "Aid should be allocated on the basis of recipient need, not of donor interest" (Neumayer, 2005).

Development policies should shift from reliance on food aid to providing farmers with the tools and resources needed to cultivate agricultural crops, produce food for their own consumption and sell the surplus. Such a shift would establish a "sustainable exit from the poverty trap, thereby decreasing the requirement for aid" (Sanchez, 2009). Appropriate governmental policies oriented towards the economy could prevent disincentive effects of food aid programs on governments and food producers of recipient countries (Bezuneh et al., 2003). Limiting food aid only to circumstances of true domestic production declines is one proposed strategy to counteract the disincentive effect of international food aid (Tadesse & Shively, 2009). Research demonstrated that donation of food aid in cash equivalent allows aid-receiving countries to improve their local economies while sourcing culturally relevant food products (Clapp, 2005).

One ton of maize grain purchased locally or from neighboring to Africa countries costs $320. It costs the U.S. $812 to deliver 1 ton of maize to Africa. As part of the Millennium Villages project, small holder farmers were granted access to improved seeds, fertilizers, technical support and markets. It costs an average of $135 to provide Millennium Village small holder farmers with improved seeds and fertilizers necessary to produce an additional ton of maize which is 6 times less than what the cost for 1 ton of food aid would cost. Maize yields increased from 1.7 to 4.1 tons per hectare as a consequence of the initiative. Therefore, "shifting 50% of the current US food-aid budget to 'smart' subsidies or credit could help millions supply their own food and meet much of the aid demand." The UN Secretary General Ban Ki-moon is coordinating initiatives designed to provide financial support to developing countries seeking agricultural investment. The Spanish government has allocated U.S. $1.3

billion over a five-year period for such initiative and the European parliament has pledged to provide a similar amount for farm investments. The prevalence of chronically hungry people is projected to begin declining simultaneously with the emergence of programs designed to merge "food aid with reliable farming investment" (Sanchez, 2009). The U.S. and other G-8 countries leaders "pledged $20 billion at the 2009 Summit to support a global effort to strengthen agriculture in developing countries" in an attempt to address the contributing to food insecurity factors (Fleige et al., 2010).

U.S. food aid would be more effective provided it adhered to international human rights guidelines, including cultural protection and respect, abstaining from creating dependency on food aid and not discriminating politically (Marchione & Messer, 2010).

Food aid is a "marginal resource" and not the means to eradicating hunger and poverty in developing countries. Reform in economic policies is necessary in order to promote economic growth and alleviate poverty and hunger (Awokuse, 2011).

19.8 GMO Food Aid

The U.S. is presently the major producer of GM crops and the main donor of food aid containing GMOs. GM food aid has been present since its commercial release in the U.S. in the mid-1990s. A "segregated system" for non-GM and GM crops in the U.S. is lacking, contributing to "widespread cross contamination." This is significant as U.S. food aid account comprises 60% of all food aid donations. Discussions on biosafety issues relative to trade of GMOs began in 1996 under the Convention on Biological Diversity, yet GMOs presence in food aid transactions were largely overlooked.

GMOs were first discovered in food aid donations in 2000. The U.S. Agency for International Development (USDA) and the WFP had been sending 3.5 million tons of GM food aid per year. "Such shipments were often in contravention of the national regulations of the recipient country." The WFP channeled U.S. GM food aid to Ecuador which was the first developing country known to receive GM food aid. The GM food aid was disposed of as a result of complaints addressed by Ecuador. U.S. GM food aid was also shipped to India and Sudan in 2000. GMO soy was discovered in U.S. food aid shipments sent to Uganda and Columbia in 2001. U.S. GM maize crops were further shipped to Bolivia in 2002 "despite that country's moratorium on the import of GMO crops." The U.S. GM food aid in Bolivia was also found to contain GM StarLink corn which was specifically approved for animal feed, "but which nonetheless managed to enter the human food system in the United States in the fall of 2000." WFP further channeled U.S. GM corn food aid to Nicaragua in 2002 which was perceived as "particularly controversial" as Nicaragua is "a center of origin for corn." Allergenicity and the contamination of wild crops with GMOs were two of the major concerns expressed by GM food aid-recipient countries. If planted rather consumed, GM food aid grains "could reduce biodiversity by contaminating and driving out local varieties. Once GMOs are released into an environment, they are difficult, if not impossible, to remove." Food aid is typically planted by farmers when it is donated in the form of grains form as farmers may have "exhausted their seed supply as food in times of crisis."

Until mid-2002, GM food aid shipments were directed towards countries that experienced food deficits, yet "were not facing acute food aid shortage." The "looming famine" in southern Africa in mid-2002 shifted such trend. The food crisis was described as "the worst food shortage faced by the region in fifty years." 14 million people in six countries in South Africa "faced imminent food shortages." Floods and droughts were two of the "imminent" factors contributing to the famine crisis in southern Africa in the mid-2002. Secondary causes included the conflict in Angola, the high prevalence of HIV/AIDS in the affected region, the domestic agricultural policies and the impact of trade liberalization.

The U.S. responded to the crisis by shipping 500,000 tons of whole maize to the affected region in southern Africa in the summer and fall of 2002. 75% of the whole maize food aid contained GMOs. Zimbabwe, Zambia, Malawi, Mozambique, Lesotho and Swaziland were the aid-recipient countries in the region. The WFP and NGOs channeled the food aid. The recipients of the food aid in South Africa were not made aware that the U.S. food aid shipment contained GMOs. A number of southern African countries refused the GM food aid due to health precautions as well as due to concerns that the GM food aid would contaminate domestic crops thus compromising future exports to European countries. Many South African countries eventually accepted the GM food aid for as long as it was milled first, yet Zambia refused the GM food aid for the general population at all cost claiming that "any health problems that might arise from eating GMOs would be too costly for the country to address." The Zambian President Levy Mwanawasa perceived GM food aid as "poison." Zambia accepted it only for the 130,000 Angolan refugees residing within its borders. The U.S. argued that it would not be able to provide non-GM food aid and did not agree to cover milling costs. WFP managed to mill the GM food aid and was able to "fortify the grain to raise its micronutrient content." The U.S. eventually provided Zambia with 300,000 tons of GM-free maize "after heavy international pressure to do so." The Zambian government authorized a scientific delegation to study GM food aid safety. The reports from the delegation, released in the fall of 2002, "cautioned against the acceptance of GMOs in Zambia, much to the disappointment of the United States." The U.S. "blamed Europe's moratorium on imports of GM foods and seed for contributing to hunger in southern Africa...Europe has not followed the lead of the United States on GM food aid policy." The EU moratorium on GM food imports has contributed to "a significant loss of markets for U.S. grain." The U.S. has lost approximately $300 million per year in sales of maize to Europe. The U.S. expressed a concern that the EU position on GMOs influenced "too many countries, including those in Africa" and "seriously" considered filing a formal complaint against the EU at the WTO. The U.S. argued that even though WTO allows countries to ban products on the basis of safety concerns, the EU had not gathered any evidence relative to GMOs in the past five years which the U.S. believed was sufficient time to do so. The U.S. "put pressure on Europe to remove its moratorium" in the fall of 2002 (i.e., before the crisis) and at the beginning of 2003, culminating with the U.S. launching a formal complaint against Europe at the WTO in May 2003. Canada and Argentina joined the formal complaint. U.S. President George W. Bush stated that: "European governments should join – not hinder – the great cause of ending hunger in Africa." Egypt which has "an active agricultural biotechnology research program" originally joined the formal complaint, but later withdrew from it as Europe is a major importer of fresh fruits and vegetables produced in Egypt. The U.S., hoping that the participation of Egypt in the formal complaint would strengthen its argument regarding the benefits of GM crops for Africa, "retaliated against Europe for its withdrawal by pulling out of talks on a U.S.-Egyptian free trade agreement." In response to the formal complaint, the EU stated: "Food aid to starving populations should be about meeting the urgent humanitarian needs of those who are in need. It should not be about trying to advance the case for GM food abroad." Sudan requires that received food aid is GM-free. The U.S. "pressured" Sudan to accept GM food aid in mid-2003. In response, Sudan issued a six-month waiver to its legislation in order to provide the U.S. with more time to supply Sudan with GM-free food aid. In March 2004, the U.S. "threatened to cut Sudan off from food aid completely" which "prompted" Sudan to extend the issued waiver to early 2005 (Clapp, 2005).

19.8.1 GMO Food Aid Legislation

The "rules" in relation to GMO trade were "unclear" locally and globally between the mid-1990s and 2003. This was the time when controversies regarding GM food aid were at their highest. Only a few developing countries had established "any" GMOs imports legislations, "let alone GMO food aid" by mid-2002. South Africa and Zimbabwe were the only sub-Saharan African countries that had established biosafety laws at that time. Zimbabwe "has not approved any GM crops for commercial release" while South Africa is the only country in sub-Saharan Africa to approve the commercial production of GM crops. A number of additional African countries began to develop GMO import legislations since mid-2002. In response to the crisis in 2002, the Southern African Development Community (SADC) issued guidelines for policy on GMOs in southern Africa, stipulating that informed consent is to be obtained from the aid-recipient country prior to delivery of GM food aid and that GM food aid shipment must be labeled. Such guidelines were, however, not available during the crisis in 2002. The WFP established policies regarding GMOs in food aid in mid-2003. The policies state that WFP will only channel GM food aid to aid-recipient countries that approve donations of GM food aid. Policies are further set in place to ensure that the channeled food aid is considered safe by both the donor and recipient countries.

"Economic and political incentives, inextricably tied to corporate interests in agricultural biotechnology, appear once again to be important factors behind the U.S. position on GM food aid in particular."

The "inability" of the U.S. to find export markets for its GM maize is perceived as one of the main reasons why the U.S. continues to give food aid in kind rather than in cash equivalent.

The U.S. Department of Agriculture (USDA) "works in

close cooperation with the agricultural biotechnology industry" and is "responsible for regulating biotechnology in the United States." The USDA "also oversees Title I food aid." The USAID is responsible for Title II and Title III food aid and also takes an active role in promoting agricultural biotechnology in the developing world via educational programs. The USAID has spent "U.S. $100 million for that purpose in recent years." On the surface, the reason for such promotional initiatives is the opening of "new markets in the future," yet critics believe that by such efforts, the U.S. is attempting to "pave the way for the introduction of pro-GM legislation to facilitate the export of GM crops and seeds around the world" (Clapp, 2005).

Discussion

Several conclusions pertaining to deceit, profit and power within the context of created problems could be drawn from the "'Non-Profit' Sector & 'Food Aid'" chapter.

First, the non-profit sector is being transformed into a for-profit sector which is a clear indication that governments prioritize profit and power over altruism. Food aid organizations represent a perfect example of governmental priorities.

Despite the availability of food, food aid organizations deny food access to certain parts of the world solely on the basis of political, geographical and military interests of the donor country. It could be therefore deduced that food access translates into donor countries' acquisition of power (i.e., creating political and military alliances with receiving country) while lack of food access dis-empowers the receiving country, automatically shifting more power to the donor country. The problem is thus manifested via the polarity of food aid access in spite of food availability for all receiving countries. Power is also manifested by the creation of dependency on food aid. Only symbolic measures are undertaken to ensure that supported by food aid countries become independent, which ensures continuation of power.

Furthermore, the formal complaint against Europe for delaying the release of GM crops as well as the denying of food aid to Sudan as soon as Sudan refuses to accept GMO food aid are clear indication that profit (i.e., through GMO) is at the heart of food aid organizations. Profit is also generated when, for example, all U.S. food aid is processed and transported on U.S. vehicles, an initiative which ensures U.S.-based profit, yet threatens the survival of thousands of people worldwide. The secrecy surrounding food aid organizations is an additional indicator that the food aid sector deceives humanity in the name of profit and power. The conclusion that can be drawn regarding food aid is that hunger and food scarcity exist because of food aid organizations, not in spite of them.

Two different examples portraying the creation of a problem (i.e., food scarcity) as a means of generating governmental and industrial profit and power pertain to the maize example from the "Salvation" chapter and the bioenergy example from "From Energy Crisis to Free Energy." Both instances illustrate that food is available, yet it is being misused. While millions of people are dying from starvation around the world, agricultural crops are used for fuel as the "'Non-Profit' Sector & 'Food Aid'" chapter also suggests. At the same time, as "Waste Management" is going to demonstrate, material that can be used as fuel is buried under the ground. The "Animal Mutilation" chapter demonstrated that animal livestock is being fed with grains. One can imagine the amount of grains used to feed meat-producing animals from roughly the moment of their birth to the moment of slaughter. While it could be deduced that the meat of a cow, for example, is fairly quickly consumed, the grains fed to that same cow during her entire lifespan of about 6 years as "Animal Mutilation" suggested, could last a human being years to consume. Such examples clearly indicate that food scarcity is purposefully orchestrated in order for governments and industries to generate profit and power under the guise of food aid organizations.

The most striking example of how problems are created in order to acquire financial gain and power is the instance with Haiti. Thousands of sandwiches were prepared by volunteers of a non-profit organization literally days before (not after) the earthquake of Haiti. When the earthquake took place, sandwiches were ready to be distributed. Given the evidence provided so far in "Sovereign Terra" of how problems are created to generate profit and power, it would not be surprising to find that the Haiti earthquake was, in actuality, manufactured. It is safe to deduce that one of the purposes of a deliberate earthquake is to portray a compassionate nature of non-profit organizations in order for humanity to continue supporting the deceit morally and financially thus increasing the profit and power acquired by "non-profit" corporations.

References

Alston, J. M., Sumner, D. A. & Vosti, S. A. (2006). Are agricultural policies making us fat? Likely links between agricultural policies and human nutrition and obesity, and their policy implications. *Review of Agricultural Economics, 28*(3), 313–322.

Andersson, N., Paredes-Solis, S., Cockcroft, A. & Sherr, L. (2012). Epidemiological assessment of food aid in the Bosnian conflict, 1994–97. *Disasters, 36*(2), 249-269.

Anekwe, T. D. & Rahkovsky, I. (2014). The association between food prices and the blood glucose level of US adults with type 2 diabetes. *American Journal of Public Health, 104*(4), 678-685.

Asfaw, A. (2006). The role of food price policy in determining the prevalence of obesity: Evidence from Egypt. *Review of Agricultural Economics, 28*(3), 305-312.

Awokuse, T. O. (2011). Food aid impacts on recipient developing countries: A review of empirical methods and evidence. *Journal of International Development, 23*, 493-514.

Ballard-Reisch, D. (2011). Feminist reflections on food aid: the case of Numana in Haiti. *Women & Language, 34*(1), 53-62.

Basu, S., McKee, M., Galea, G. & Stuckler, D. (2013). Relationship of soft drink consumption to global overweight, obesity, and diabetes: A cross-national analysis of 75 countries. *American Journal of Public Health, 103*(11), 2071-2077.

Bezuneh, M., Deaton, B. & Zuhair, S. (2003). Food Aid disincentives: The Tunisian experience. *Review of Development Economics, 7*(4), 609-621.

Bliss, R. M. (2011). Fully cooked Emergency Aid food. *Agricultural Research, 59*(7), 4-7.

Castetbon, K., Mejean, C., Deschamps, V., Bellin-Lestienne, C., Oleko, A., Darmon, N. & Hercberg, S. (2011). Dietary behaviour and nutritional status in underprivileged people using food aid (ABENA study, 2004-2005). *Journal of Human Nutrition and Dietetics, 24*, 560-571.

Clapp, J. (2005). The political economy of food aid in an era of agricultural biotechnology. *Global Governance, 11*, 467-485.

Conceicao, P. & Mendoza, R. U. (2009). Anatomy of the global food crisis. *Third World Quarterly, 30*, 1159-1182.

Dodge, N. (2013). Effect of climate change and food insecurity on low-income households. *American Journal of Public Health, 103*(1), e4.

Essex, J. (2012). Idle hands are the devil's tools: The geopolitics and geoeconomics of hunger. *Annals of the Association of American Geographers, 102*(1), 191-207.

Fleige, L. E., Moore, W. R., Garlick, P. J., Murphy, S. P., Turner, E. H., Dunn, M. L., van Lengerich, B., Orthoefer, F. T. & Schaefer, S. E. (2010). Recommendations for optimization of fortified and blended food aid products from the United States. *Nutrition Reviews, 68*(5), 290-315.

Grasso, J. (2003). The politics of food aid: John, F. Kennedy and famine in China. *Diplomacy & Statecraft, 14*(4), 153-178.

Husband, A. (2013). Climate change and the role of food price in determining obesity risk. *American Journal of Public Health, 103*(1), e2.

Hoddinott, J., Cohen, M. J. & Barrett, C. B. (2008). Renegotiating the food aid convention: Background, context, and issues. *Global Governance, 14*, 283-304.

Jager, U. P. & Rehli, F. (2012). Cooperative power relations between nonprofit board chairs and executive directors. *Nonprofit Management & Leadership, 23*(2), 219-236.

Jordan, C. F. (2002). Genetic engineering, the farm crisis, and world hunger. *BioScience, 52*(6), 523-528.

Kabbani, N. S. & Kmeid, M. Y. (2005). The role of food assistance in helping food insecure households escape hunger. *Review of Agricultural Economics, 27*(3), 439-445.

Kasper, J., Gupta, S. K., Tran, P., Cook J. T. & Meyers, A. E. (2000). Hunger in legal immigrants in California, Texas, and Illinois. *American Journal of Public Health, 90*(10), 1629-1633.

Kirwan, B. E. & McMillan, M. (2007). Food aid and poverty. *American Journal of Agricultural Economics, 89*(5), 1152-1160.

Kneteman, C. (2009). Tied food aid: Export subsidy in the guise of charity. *Third World Quarterly, 30*(6), 1215-1225.

Kuhlgatz, C. & Abdulai, A. (2012). Food aid and malnutrition in developing countries: Evidence from global food aid allocation. *Journal of Development Studies, 48*(12), 1765-1783.

Lentz, E. C. & Barrett, C. B. (2014). The negligible welfare effects of the international food aid provisions in the 2014 Farm Bill. *Choices: The Magazine of Food, Farm & Resource Issues, 29*(3), 1-5.

LeRoux, K. & Feeney, M. K. (2013). Factors attracting individuals to nonprofit management over public and private sector management. *Nonprofit Management & Leadership,* 24(1), 43-62.

Marchione, T. J. & Messer, E. (2010). Food aid and the world hunger solution: Why the U.S. should use a human rights approach. *Food and Foodways, 18*, 10-27.

Margulis, M. E. (2013). The Reginrie Connplex for food security: Implications for the global hunger challenge. *Global Governance, 19*, 53-67.

Naughton-Treves, L., Holland, M. B. & Brandon, K. (2005). The role of protected areas in conserving biodiversity and sustaining local livelihoods. *Annual Review of Environment & Resources, 30*, 219-252.

Neumayer, E. (2005). Is the allocation of food aid free from donor interest bias? *The Journal of Development Studies, 41*(3), 394-411.

Pennerstorfer, A. & Trukeschitz, B. (2012). Voluntary contributions and wages in nonprofit organizations. *Nonprofit Management & Leadership, 23*(2), 181-191.

Perez-Exposito, A. B. & Klein, B. P. (2009). Impact of fortified blended food aid products on nutritional status of infants and young children in developing countries. *Nutrition Reviews, 67*(12), 706-718.

Porter, C. (2010). Safety nets or investment in the future: Does food aid have any long-term impact on children's growth? *Journal of International Development, 22*, 1134-1145.

Rylander, C., Sandanger, T. M., Engeset, D., Lund, E. (2014). Consumption of lean fish reduces the risk of Type 2 Diabetes Mellitus: A prospective population based cohort study of Norwegian women. *PLoS ONE, 9*(2), 1-10.

Sanchez, P. A. (2009). A smarter way to combat hunger. *Nature, 458*, 148.

Schmidt, M. R. & Wei, W. (2006). Loss of agro-biodiversity, uncertainty, and perceived control: A comparative risk perception study in Austria and China. *Risk Analysis, 26*(2), 455-470.

Svedberg, P. (1978). World food sufficiency and meat consumption. *American Journal of Agricultural Economics, 60*(4), 661-666.

Tadesse, G. & Shively, G. (2009). Food Aid, food prices, and producer disincentives in Ethiopia. *American Journal of Agricultural Economics, 91*(4), 942-955.

Toft, K. H. (2012). GMOs and global justice: Applying global justice theory to the case of genetically modified crops and food. *Journal of Agricultural & Environmental Ethics,* 25, 223-237.

Tucker, B. P. & Parker, L. D. (2013). Managerial control and strategy in nonprofit organizations doing the right things for the wrong reasons? *Nonprofit Management & Leadership, 24*(1), 87-107.

Van Puyvelde, S., Caers, R., Du Bois, C. & Jegers, M. (2013). Agency problems between managers and employees in nonprofit organizations: A discrete choice experiment. *Nonprofit Management & Leadership, 24*(1), 63-85.

Wehler, C., Weinreb, L. F., Huntington, N., Scott, R., Hosmer, D., Fletcher, K., Goldberg, R. & Gundersen, G. (2004). Risk and protective factors for adult and child hunger among low-income housed and homeless female-headed families. *American Journal of Public Health, 94*(1), 109-115.

Chapter 20

OIL SOLICIT INDUSTRY

Overview

Petroleum availability, use, peak and pollution are discussed. Oil drilling examples, policies, regulations and catastrophes are further presented. Examples of human rights and environmental violations pertaining to the oil industry are elaborated upon. "Oil Solicit Industry" concludes by elaborating on the BP Deepwater Horizon disaster.

20.1 Petroleum

Petroleum is a non-renewable resource. Prior to the global crisis from 2007 until 2009, 85 million barrels of petroleum were used per day on a global scale. In 2008, the United States (U.S.) used about 19.5 million barrels of oil per day, 70% of which were allocated to transportation use. 94% of all energy used in the transportation sector was derived from petroleum in 2008. Petroleum has additional applications including heating, operating machinery, plastics and pesticides. The U.S. imports nearly 60% of the petroleum it uses, which makes the country as the world's largest importer of petroleum (Schwartz et al., 2011). Currently, 33% of petroleum produced in the U.S. is derived from fracking (Howarth & Engelder, 2011).

20.1.1 Peak

"The world petroleum production peak is imminent, and we are entering an unprecedented era of petroleum scarcity." The peak of petroleum production is projected to take place between the first and third decade of the 21st century and begin to decline thereafter. It is challenging to estimate the exact timing of the petroleum peak due to "absence of reliable information on past production and reserves in many oil fields." Global production is expected to decrease by 2% to 3% per year after the petroleum peak has been reached. The gap between petroleum demand and supply is increasing. When the prices of oil increase substantially, humanity will most likely focus on policy options that would contribute to further environmental degradation worldwide and severe human health outcomes. Petroleum geologists expressed concerns regarding the geological limitations of oil production more than 5 decades ago (Schwartz et al., 2011).

20.1.2 Petroleum-Contaminated Soils

U.S. cleanup standards for petroleum-contaminated soils (PCS) at the state level were established in the mid-1980s. A 1985 survey revealed that a total of 22 out of 50 states had established PCS cleanup levels. 5 states had established formal cleanup levels while the remaining states had established informal cleanup levels. 8 states considered establishing formal PCS cleanup levels (Kostecki et al., 2001).

Large quantities of crude oil and oil particles were discharged in desert soil in Kuwait as a result of the invasion and occupation of Iraq over Kuwait. About 70 oil lakes containing 9 million m³ of oil reaching approximately 0.7 m and in a few cases 2.5 m of depth were formed, spreading over an area of approximately 50 km². It was estimated that more than 20 million m³ of heavily polluted with oil soil would require treatment. Substantial amounts of severely contaminated soil would need to be remedied in order for the soil to be suitable for agricultural and animal production. Bioremediation is evaluated as the most cost-effective method to effectively treat heavily contaminated soil in Kuwait (Al-Daher et al., 2001).

20.2 Oil Drilling

20.2.1 Great Lakes

The Great Lakes represent the largest surface freshwater resource in the world. Fresh surface water extends to 5440 cubic miles. The Great Lakes is composed of five Great Lakes – Lake Michigan, Superior, Huron, Erie and Ontario. Along with connecting channels and St. Lawrence River, the Great Lakes represent 95% of surface freshwater in the U.S. and 20% of worldwide supply. The Great Lakes supply millions of Americans and Canadians with fresh drinking water. Canada, the U.S. and 8 U.S. states are responsible for the regulatory jurisdiction of the Great Lakes.

The regulatory jurisdiction regime is inconsistent, lacks uniformity and provides the basis for possible trans-boundary pollution. Great Lakes management of one country would most likely have "spillover consequences" for the neighboring country. The Great Lakes contain substantial oil and gas resources. Oil drilling in the Great Lakes is mostly banned. There is a robust public support for protection of the Great Lakes. Drilling for oil and gas in the Great Lakes would pose risks to the freshwater resources of the Great Lakes. In 1985, all 8 U.S. Great Lakes states signed a statement, opposing oil drilling in the Great Lakes. Following the signing of the statement, some states prohibited oil and gas drilling in the Great Lakes, others only prohibited oil drilling and several states had no explicit laws pertaining to oil and gas drilling in the Great Lakes. Congress indefinitely banned drilling in 2005 following a report outlining the economic, environmental and navigational impacts of oil and gas drilling in the Great Lakes. Section 386 of the Energy Policy Act of 2005 states: "[n]o Federal or State permit or lease shall be issued for new oil and gas slant, directional, or offshore drilling in or under one or more of the Great Lakes" (Hall, 2011).

Despite the fact that the U.S. has banned drilling in the

Great Lakes, "the potential remains for Canadian drilling operations to pollute this important freshwater resource" (der Mude, A. V., 2011). Ontario, Canada, is the only province in Canada which has jurisdiction over the Great Lakes. Ontario has not banned oil and gas drilling in the Great Lakes. "The Canadian side of the Great Lakes is still open to drilling...So while Ontario continues to allow offshore gas wells and directional drilling of oil wells in the Great Lakes, it is a very small industry with minimal economic and energy supply value." Given the significance of protecting the surface freshwater system of the Great Lakes (both in Canada and in the U.S.) and the relatively minor energy and economic value derived from oil and gas Great Lakes drilling in Canada, it is necessary and logical to establish an international agreement to protect the surface freshwater system in the Great Lakes (Hall, 2011).

20.2.2 The Arctic

The Arctic, home of wildlife and indigenous populations, is estimated to contain the largest reserves of natural gas and oil in the world. Until recently, the Arctic was the last non-industrialized region in the world. "In 2010, the EPA approved two permits for Shell to begin offshore exploratory drilling in the Arctic's Chukchi and Beaufort Seas with the drill-ship Discoverer." The CAA requires The U.S. EPA to preserve the environment and the air quality of pristine areas sacred to human beings and to "protect the public welfare from the adverse effects of ruthless industrial expansion." Drilling operations endanger the Arctic's ecosystem by releasing GHG emissions, hence contributing to global climate change. Permitting Shell exploratory drilling in the Arctic would contribute to hundreds of tons of air pollutants being emitted in the atmosphere each year. The support vessels of Discoverer are expected to emit twenty tons of PM per year, most of which in the form of black carbon. More than one ton

314

of nitrogen oxide is projected to be emitted per day for a supply vessel that is not connected to Discoverer, but uses its own engines. EPA allowed the emission of high levels of air pollution without the use of best available control technology (BACT) for support ships thus threatening the pristine state of the environment at the Arctic. BACT is only applicable to vessels physically connected to the main drill-ship, Discoverer. Yet, support vessels are not physically attached to Discoverer and are therefore exempt of BACT requirements. The U.S. Court of Appeals for the Ninth Circuit supported EPA's legal procedures and decisions, illustrating "courts' limited ability to question agency decisions that run counter to congressional initiatives to protect the environment" (Warren, 2014).

20.2.3 Regulation

20.2.3.1 Outer Continental Shelf Lands Act (OCSLA)

The Outer Continental Shelf Lands Act (OCSLA) governs the development of gas and oil resources owned by the federal government on the Outer Continental Shelf. OCSLA incorporates few standards and mandates designed to ensure protection of the environment, health and safety. The penalty provisions under the Act are "extremely modest...at a level unlikely to deter risky conduct." The Obama Administration and Congress proposed pertinent reforms to the OCSLA such as the establishment of more "rigorous standards for BOPs," improved enforcement of existing law and more robust consideration of environmental impact. The Consolidated Land Energy and Aquatic Resources (CLEAR) Act incorporated proposed reforms. While the bill passed the House of Representatives, it "stalled in the Senate and will certainly not be revived in the current Congress" (Flournoy, 2011).

20.2.3.2 Minerals Management Service's (MMS)

The Minerals Management Service's (MMS) regulatory approach under OCSLA "relied heavily on standards developed by and voluntarily agreed to by industry." MMS, now BOEMRE, has lacked funding and adequate resources such as staff and inspection resources, preventing effective regulation of offshore oil exploration and drilling activities. Between 1982 and 2007, MMS staffing resources decreased 36%. Inspections are described as "infrequent, rarely announced," consisting primarily of verifying paperwork while inspectors are not provided with most fundamental equipment such as laptops (Flournoy, 2011).

20.2.3.3 Oil Pollution Act of 1990 (OPA)

The Oil Pollution Act of 1990 (OPA) is responsible for the establishment of liability relative to oil discharge or the possibility of oil discharge into navigable U.S. waters. A different OPA objective is the restoration of compromised by oil spills natural resources. "Until recently, holding parties liable for damages to natural resources was inconceivable." Even with the passage of provisions to natural resources as part of the Clean Water Act, "little effort was made to calculate the value of various natural resources" (Kornfeld, 2011).

20.3 Oil Industry

20.3.1 Oil Companies

A 2005 Gallup Poll survey revealed that "big business" was the most commonly distrusted U.S. institution (Watts, 2005). "The wealth of the large oil companies is almost impossible to conceive." The five major oil companies are Shell, BP, Chevron, ExxonMobil and Conoco Phillips. Combined, they generated a profit of $100 billion in 2008. BP reported a profit of $1.785 billion for the third quarter of 2010 despite $40 billion liability estimates from the BP Deepwater Horizon disaster. Such wealth is "a powerful force" that is deployed by the oil and gas industry to "influence both the composition of the legislature and the legislation that emerges from Congress". 2010 campaign contributions from the oil and gas industry only exceeded $23 million while 2008 total campaign contributions accounted to $35 million. The five major oil companies contributed more than $92 million for lobbying in 2009 while the industry as a whole contributed over $175 million. The oil and gas industry employed 781 lobbyists in 2009 and 744 in 2010, two-thirds of whom "had previously worked for the federal government." The oil and gas industries "have opposed, and will continue to oppose, efforts at meaningful reform" (Flournoy, 2011).

20.3.2 Hollow Government

In *Citizens United vs. Federal Election Commission*, the Supreme Court made a decision that "corporations are entitled to the same political speech rights as individuals" which "has increased the power of corporate interests." The last 3 decades have been characterized by "an incessant drumbeat" for smaller government, less regulations, and cuts in federal spending. The result of such an "ideology" is the so called "hollow government." The BP Deepwater Horizon disaster can be perceived as a consequence of such "anti-government ideology." Such ideology "focuses on government as the threat to the public, and ignores the power and political influence that economic interests wield, often to the detriment of public health and safety, and the environment. This in turn shields the tremendous power of corporate interests from public view and makes it unlikely that the public will uncover the reality of and threat posed by hollow government." Furthermore, "the current energy policy provides hefty subsidies for the highly profitable oil and gas industries to continue with their unwavering focus on producing more oil and gas" (Flournoy,

2011).

20.3.3 U.S. Energy Policy

One component of U.S. energy policy focuses on privatization of public natural resources. U.S. energy policy is designed to sell natural resources at "bargain prices" presumably as a means to promote development. U.S. royalty rates were compared to the royalty rates of 103 other jurisdictions. Findings revealed that 11 jurisdictions had royalty rates lower than the royalty rates of the U.S. The current energy policy "strongly encourages all-out exploitation of remaining domestic fossil fuel resources, and deepwater oil reserves in particular. If the public and elected officials believe that the risks that produced the Macondo Well blowout are unacceptable, an energy policy that will move us towards a clean energy path is a logical response" (Flournoy, 2011).

20.3.4 Violations

The last 20 years have been marked by "an aggressive push by corporations to expand their operations into all corners of the world...operating with, and through, all manner of undemocratic governments...and...the rise of a global human rights movement...for whom corporate business practices...must be made transparent and accountable in terms of a broad interpretation of human rights" (Watts, 2005).

20.3.4.1 Texaco Inc.

In 1993, a Philadelphia law firm filed a $1.5 billion suit against Texaco Inc. on behalf of 30,000 citizens of Ecuador. 20 years of oil drilling on the Oriente contributed to water contamination, the development of illnesses and ecological deterioration. Irresponsible corporate oil operations endangering the lives of thousands of people were at the heart of allegations. A U.S.

corporation was allegedly violating international law by contributing to environmental deterioration abroad (Watts, 2005).

20.3.4.2 Unocal

The NGO, Earthrights International, filed a suit against Unocal arguing that the Yadama pipeline had directly contributed to a number of human rights abuses such as "rape, death, and the disruption of a local way of life" (Watts, 2005).

20.3.4.3 Iraq

The Cheney Energy Task Force report of 2001 "heated public and scientific debate over oil scarcity and Hubbert's peak." The U.S. government was at war with Iraq within months of the release of the report which spurred a debate as to whether this was "blood for oil...over the operations of the oil majors in the reconstruction of Iraq."

     The petro-state and oil complex have spurred a substantial amount of research devoted to the examination of the association between oil, conflict, violence and antidemocratic politics. "Right from the start in the early twentieth century, oil extraction has gone along with the most ruthless and open imperial violence, with repeated warfare..., and with a sort of lawlessness characteristics of the corporate frontier. Iraq is the result of just these processes." "Blood for oil" is perceived as the primary motive for the occupation of Mesopotamia as part of the...war in Iraq. In 1944, the U.S. State Department referred to oil as a "stupendous source of strategic power" (Watts, 2005).

20.3.5 Developing World

Oil operations in the developing world exist as a joint enterprise involving governments, service companies, independents and supermajors. All involved parties endeavor to shift responsibility and blame to one another and to other parties. Generally, oil operations of all involved parties are embedded in secrecy "typically couched in terms of national security" thus hindering verification of actual oil operations. National oil companies, including Iranian National Oil and Saudi Arabian Oil are described as "black boxes" as what is known about them is extremely scarce (Watts, 2005).

20.3.6 Corruption

Corruption pertaining to the oil sector operates on various levels, including the privatization of public office by state officials within the oil sector, the theft of oil revenues by political and military representatives, payoffs for varied contractual tasks, unlawful commissions and illegal "oil-for-arms deals". In a number of oil states, "the oil portfolio is under the direct jurisdiction of the president." "This looting of the state, in turn, worsens income distribution, takes recourses for development and social welfare programs, and represents a radical (and illegal) privatization of a national resource" (Watts, 2005).

20.3.7 Indigenous People

Often, oil operations take place within the land of indigenous people. "Companies often only pay lip service to local community" while offering "irregular and minimal payments for the use of tribal and other lands...The entire arena of land is deeply fraught" (Watts, 2005).

20.3.8 Human Rights

In 2003, Human Rights Watch conducted a study concerning the "complicity of oil companies in human rights violations in Sudan." Such "clashes" between the oil industry and human rights advocacy groups "have become increasingly commonplace." The oil industry is "distinguished by its size, power, and strategic significance, but also by a long association with violence and empire."

"The complex relations between oil and violence mean that many extractive industries operate in conditions of deep enmity and conflict, and occasionally civil war and insurgency." Human rights concerns take place due to the fact that ongoing oil operations could contribute to continued or enhanced military action. Indonesia imposed martial law and victimized Aceh in 1990, granting the military sector the responsibility of planning and carrying out oil and gas installations. The role of oil companies in Afghanistan relative to the rise of Taliban movement and the civil war was considered "deeply unethical." A Human Rights Watch report on Sudan further reveals that the Sudanese Muslim government established a cordon sanitaire around a 936-mile oil pipeline, transporting oil from South Sudan to North Sudan. Since 1999, the Sudanese Muslim government took an active part in "massive human rights violations, moving people from the oil-producing regions - 175,000 remain displaced - with the complicity of the companies." For instance, due to conflicts, Lundin Oil ceased oil operations at an oil reserve it had discovered at Thar Jath in 1999. Lundin Oil was able to resume oil operations following "aerial bombing and executions by state forces." The Talisman Oil Company also took part in the civil war, "with evidence of the deployment of mujahideen and child combatants to confer security around Block 5." A small number of the oil companies "spoke out against the atrocities or made any effort to provide community assistance...and a number allowed the government to use their facilities during its military activities against the pastoral communities" (Watts, 2005).

20.3.9 Environmental Degradation

Environmental degradation related to oil operations results essentially from blowouts, oil spills, hydrocarbon releases and the outcomes of gas flaring. A number of oil spills are neither documented nor intervened upon. "Company cleanup is often tardy, limited, and shrouded in secrecy to prevent accurate estimation of spillage and damage." Often, oil companies do not act upon uncovering the impact of oil operations on the environment. Livelihoods can be affected by environmental pollution, explosions, and displacement. Nigeria has the highest rate of gas flaring in the world – 12% of global gas flaring in 2002 (Watts, 2005).

20.3.10 Security

"Oil companies often operate in circumstances of (a) civil war and military insurgencies..., and occasionally (b) interstate conflicts..., and (c) military governments or undemocratic regimes in which the security and military apparatuses defend or secure oil operations." Oil and gas operations are almost always secured and defended by a combination of private, state or foreign security forces. It is not accidental that invariably all foreign oil companies "operate out of highly defended paramilitary compounds and that any oil installation will have police and military posted outside their facilities." Oil companies are entitled to protecting their facilities. All oil companies employ "supernumerary police." Security arrangements between oil companies and governments are unavoidable as oil companies are often times legally mandated to report all security matters to the government. "The question of codes of conduct, internal systems of deployment, and the secrecy of such arrangements are nonetheless imperative (the security guidelines in the memoranda of understanding between governments and companies are never made public)." The U.S. military assistance to the Columbian government to protect oil pipelines "has become part of a brutal military campaign to suppress insurgents (and indigenous opposition to the oil industry)." Shell has a substantial monitoring and surveillance system of conflicts and protests in Niger Delta, Nigeria. Oil companies in Africa "have admitted importing arms and deploying private military forces, and they have been directly involved in a number of killings of oil protesters" (Watts, 2005).

20.3.11 Theft

Small-scale (tapping of fuel pipelines) and large-scale ("bunkering" via oil flow stations) oil theft represents a facet of the oil complex. 10% of U.S. oil import is derived from bunkering while 200,000 – 250,000 barrels of oil are stolen every day in Nigeria (Watts, 2005).

20.3.12 Oil Spills

A large number of oil spill events have taken place in the aquatic environment of Earth. Some of the major disasters include the Exxon Valdez oil spill of 37,000 tons of oil in Alaska, the Prestige spill of 63,000 tons offshore of Spain and the Erika spill of 20,000 tons in the South Atlantic (Kassaify et al., 2009). There have been over 4,800 oil spills in Nigeria between 1970 and 2000 (Watts, 2005).

Oil spills could have ecological and social impacts. Social relationships, human physical (i.e., acute and chronic health effects from exposure to oil) and mental health (i.e., depression and stress as a result of oil spills or as a response to oil spill), infrastructure, governance, political and economic systems could be affected by oil spill events (Lord et al., 2012).

20.3.12.1 Buzzards Bay, Massachusetts (MA)

In April 27, 2003, the commercial tanker Bouchard-120 spilled approximately 320 tons of oil in Buzzards Bay, Massachusetts (MA), U.S. Buzzards Bay is one of MA's most profitable fishing region. The oil extended over 90 miles of shoreline within the next couple of days following the spill. "Adverse weather conditions" prevented timely emergency response to the oil spill. The spill affected natural resources, including birds, shorelines, beaches, shellfish beds and salt marshes. The emergency response period of the cleanup was finalized after approximately four months in September 2003 (Lord et al., 2012).

20.3.12.2 Lebanon

In July 2006, "a war-induced" oil spill in Lebanon caused the release of 15,000 tons of heavy oil into the Eastern Mediterranean Sea. Heavy oil contains Polycyclic aromatic hydrocarbon (PAH). Study findings revealed that oysters affected by the oil spill contained certain bacterial groups at levels higher than the recommended by international standards. The high bacteriological pollution levels of oysters could pose human health risks as a great number of people worldwide consume oysters raw (Kassaify et al., 2009).

20.3.12.3 BP Deepwater Horizon

Resulting "from America's insatiable thirst for oil," the BP Deepwater Horizon disaster is "one of the worst man-made environmental disasters that the United States has experienced to date" (Kornfeld, 2011). The Macondo well is located on federally owned land 50 miles off the coast of Louisiana (Doremus, 2011). BP had drilled the Macondo oil well in the Gulf of Mexico to its "final depth" of over 18,000 feet and were "cementing the well's steel casing" when "buildup of methane gas rose to the surface," causing the destruction, collapse and sink of the Deepwater Horizon drilling rig (Flournoy, 2011). The BP Deepwater Horizon "burst into flames" on April 20, 2010 (der Mude, A. V., 2011). 11 workers were never found and 17 suffered injuries. Oil began spilling into the Gulf of Mexico. "Repeated efforts to trigger the blowout preventer failed, leaving the well to gush oil, uncontrolled." The blowout preventer (BOP) is "designed to shut off the well in the event of an imminent blowout" (Flournoy, 2011). The BP Deepwater Horizon disaster contributed to the release of 53,000 barrels of oil per day for a total of 4.9 million barrels of oil spilled into the Gulf of Mexico for three months (Kornfeld, 2011) before the well was sealed. Both technological as well as regulatory failures contributed to the BP Deepwater Horizon disaster (Flournoy, 2011).

As details regarding the BP Deepwater Horizon "became available, they revealed frustrating parallels to the 1989 Exxon Valdez oil spill in the Gulf of Alaska in terms of causation and impaired response capability." The "systemic flaws" pertinent to the oil spill in the Gulf of Alaska "were largely cloaked behind the figure of a captain with a drinking problem." The oil production governance consists of the oil industry, "generating jobs, technology, wealth, and political power" and state and federal regulatory agencies, responsible for "monitoring the industry and protecting the public from industry's market failure externalities" (Plater, 2011).

20.3.12.3.1 Regulation

Effort to examine the chain of events preceding the Deepwater Horizon disaster lead to "discouraging conclusions" regarding the oil production regulatory system in the U.S. It was "demonstrated that the MMS was a captive of the U.S. oil industry" (Steinzor, 2011).

"In 2006, in the wake of rising oil prices, legislation opened up large areas of the Gulf of Mexico to offshore oil drilling." MMS defined blowout events as "rare events and of short duration" and stated that "a subsurface blowout would have a negligible effect on [Gulf of Mexico] fish resources or commercial fishing." As part of the well exploration plan, BP refused to take into consideration the effects of a potential substantial blowout. "MMS considered oil drilling in the Gulf of Mexico to pose so little risk to the environment that it was not even required to assess that risk" (Barsa & Dana, 2011). A number of external reviews of the environmental analysis conducted by MMS were undertaken, yet "none uncovered MMS's wildly incorrect estimates of the probability, magnitude, and consequences of a blowout" relative to deepwater drilling at the Macondo well (Doremus, 2011). "Perhaps most damning is evidence that the agency actually altered environmental review documents in order to speed up approval of oil drilling" (Barsa & Dana, 2011).

MMS received numerous reports regarding the effectiveness and reliability of blind shear rams, which were supposedly the last "failsafe" possibility to prevent a blowout. "Even where safety testing was required by regulation, it was not performed." BP, for example, never submitted test data verifying the reliability and effectiveness of the blind shear rams to be used as part of its drilling operations at Macondo well. Frank Patton, the engineer in the New Orleans MMS office who was tasked to review the BP application, did not ask for blowout protector test data statement. Frank Patton who worked for MMS and the oil industry for almost 3 decades, testified that he was never trained to look for such statement as far as he can recall. He adds that he had approved hundreds of oil drilling permits in the gulf and never required proof that the blind shear ram technology was in an operational state.

To sum, "the review process was lax, categorical exclusions proliferated, risks were either ignored or excised from review documents, and permits were approved without the required assurance" (Barsa & Dana, 2011).

The National Environmental Policy Act (NEPA), the Endangered Species Act (ESA) and the Coastal Zone Management Act (CZMA) are responsible for assisting the public in understanding the possible environmental trade-offs of offshore energy development decisions prior to committing to them and for ensuring that such trade-offs remain within acceptable levels. The BP Deepwater Horizon catastrophe raised important questions regarding the "ability of those laws to fulfill their intended purpose." The Department of Interior demonstrates scarce interest in learning any lessons pertaining to environmental review (Doremus, 2011).

The Secretary of the Interior, Kenneth Salazar fired the head of the MMS and restructured the organization not long after the BP Deepwater Horizon disaster became "evident". The federal government has undertaken certain actions to address the catastrophe such as the adopting of updating regulations pertinent to oil and gas activities. Yet, "there are legitimate reasons to doubt" whether such initiatives are sufficient to significantly reduce the risk of another oil spill disaster (Latham, 2011).

Shortly after the disaster, "the benighted agency," MMS, was renamed the Bureau of Ocean Energy Management, Regulation and Enforcement (BOEMRE). Such conversion did not contribute to more effective regulatory mechanisms pertaining to the over 3,500 oil platforms and drilling rigs in the Gulf of Mexico (Steinzor, 2011). Creating BOEMRE in the place of MMS "appears to have done little to address the lack of resources that are critical for any agency to serve in a robust regulatory capacity" (Latham, 2011).

20.3.12.3.2 British Oil Production Regulatory System

The Deepwater Horizon oil spill in the Gulf of Mexico has prompted a search for regulatory strategies that would contribute to the prevention of such catastrophes in the future. The new Bureau of Ocean Energy Management, Regulation and Enforcement has been pressured to adopt the British regulatory safety case regime. The British safety case system requires a several hundred pages long safety plan for each specific facility. British safety cases are "strictly confidential." Only certain officials, such as company officials, top level management, regulators and, under certain circumstances, worker representatives, are permitted to see the entire plan. The present Article argues that the British safety case system should not be adopted in the U.S. as the degree of confidentiality in addition to the "levels of risk tolerated by the British system" are in conflict with "the spirit and the letter of American law."

The flaws of the British oil production regulatory system were identified as the contributing to the British Piper Alpha oil spill in 1988. The oil production facility was inspected in June 1987 and again in June 1988 - "only weeks before it blew into pieces." An aftermath report, prepared by Lord William Douglas Cullen, concluded that "those inspections were...superficial to the point of being of little use as a test of safety on the platform..." The report further highlights the shifting of responsibility for the regulation of offshore safety hazards from the British Department of Energy, considered to be the "oil industry 'sponsoring' department," to the HSE. HSE is the English equivalent to the U.S. Occupational Safety and Health Administration (OSHA) (Steinzor, 2011).

20.3.12.3.3 Aftermath

The Deepwater Horizon oil spill contributed to economic losses pertaining to fishing and tourism, chronic and acute health effects, loss of confidence in governmental institutions, social and household conflict, loss of identity, anxiety and depression (Lord et al., 2012). Thousands of birds, fish, turtles and marshlands "were left to die" as a result of the spill. Nature requires an extensive period of time to "heal itself." The President of Union Oil Company, Fred L. Hartley commented: "I don't like to call it a disaster, because there has been no loss of human life. I am amazed at the publicity for the loss of a few birds." Congressman Young further stated in June 2010: "This is not an environmental disaster, and I will say that again and again because it is a natural phenomenon. Oil has seeped into this ocean for centuries, will continue to do it. During World War II there was over 10 million barrels of oil spill from ships, and no natural catastrophe...We will lose some birds, we will lose some fixed sealife, but overall it will recover." Mr. Young won the November 2010 midterm election seat (Kornfeld, 2011).

20.3.12.3.4 Media

22% of U.S. news was devoted to the oil spill 100 days following the BP Deepwater Horizon disaster that took place in April 20, 2010 in the Gulf of Mexico. As control over the spill was gained by authorities and fishing waters in the gulf were largely re-opened on November 15, 2010, news coverage subsided to a large degree.

315 news articles concerning seafood safety from the year following the disaster were examined. About one third of articles specifically concerned risks associated with seafood consumption as a standalone topic rather in conjunction with economic or environmental risks. Government sources were the most frequent, focusing on "reassuring as to seafood safety...even from the beginning when there was little information available about actual risk." It was found that media coverage often failed to address a public health perspective, "providing inadequate context and information about different levels of prevention." Furthermore, "striking differences existed in how seafood safety messages were framed by different quoted stakeholders." Newspaper readers were likely exposed to "different risk messages" relative to their residence and depending on whether they read local or national newspapers. Seafood safety messages as a standalone topic comprised about one third of all local and national media coverage headlines. Health risks were typically paired with environmental and economic risks as well. "Local reporting of seafood risk becomes a potentially delicate

matter of gauging health and economic impacts." While health safety is imperative to communicate, disproportionately magnifying concerns could hinder seafood consumption. "Amid a complex disaster, journalists were able to present many perceptions of risk and highlight controversy without making a spectacle of seafood safety."

16% of the seafood supply in the U.S. is derived from the Gulf of Mexico. The consumption of seafood contaminated with oil increases exposure to carcinogenic PAHs. Findings from 2 studies revealed that following the disaster, human health risk from consuming oil-contaminated Gulf of Mexico seafood was low (Greiner et al., 2013).

"Cunnings suppress the accessible wealth and easy style."

By Sosé Gjelaj

Discussion

"Oil Solicit Industry" provides an ideal example of how governments and industries are quite literally partners in actual crime; one entity disguised under a different mask. Formal laws are established for humanity and not for governmental and industrial officials. Deceit, secrecy, violence, murder and environmental deterioration under the pretence of national security (i.e., problem) and in the name of profit and power are the only "laws" that governments and industries appear to adhere to.

References

Al-Daher, R., Al-Awadhi, N., Yateem, A., Balba, M. T. & ElNawawy, A. (2001). Compost soil piles for treatment of oil-contaminated soil. *Soil and Sediment Contamination, 10*(2), 197-209.

Barsa, M. & Dana, D. A. (2011). Reconceptualizing NEPA to avoid the next preventable disaster. *Boston College Environmental Affairs Law Review, 38*, 219-245.

der Mude, A. V. (2011). Endocrine-disrupting chemicals: Testing to protect future generations. *Environmental Affairs, 38*, 509-535.

Doremus, H. (2011). Through another's eyes: Getting the benefit of outside perspectives in environmental review. *Boston College Environmental Affairs Law Review, 38*, 247-280.

Flournoy, A. C. (2011). Three meta-lessons government and industry should learn from the BP Deepwater Horizon disaster and why they will not. *Boston College Environmental Affairs Law Review, 38*. 281-303.

Greiner, A. L., Lagasse, L. P., Neff, R. A., Love, D. C., Chase, R., Sokol, N. & Smith, K. C. (2013). Reassuring or risky: The presentation of seafood safety in the aftermath of the British petroleum Deepwater Horizon oil spill. *American Journal of Public Health*, e1-e9.

Hall, N. D. (2011). Oil and freshwater don't mix: transnational regulation of drilling in the Great Lakes. *Boston College Environmental Affairs Law Review, 38*, 305-316.

Howarth, R. W. & Engelder, T. (2011). Should fracking stop? *Nature, 477*, 271-275.

Kassaify, Z. G., El Hajj, R. H., Hamadeh, S. K., Zurayk, R. & Barbour, E. K. (2009). Impact of oil spill in the Mediterranean Sea on biodiversified bacteria in oysters. *Journal of Coastal Research, 25*(2), 469-473.

Kornfeld, I. E. (2011). Of dead pelicans, turtles, and marshes: Natural resources damages in the wake of the BP Deepwater Horizon spill. *Boston College Environmental Affairs Law Review, 38*, 317-342.

Kostecki, P. T., Calabrese, E. & Simmons, K. (2001). Survey of States' 2000 soils cleanup standards for petroleum contamination. *Soil and Sediment Contamination, 10*(2), 117-196.

Latham, M. A. (2011). Five thousand feet and below: The failure to adequately regulate deepwater oil production technology. *Boston College Environmental Affairs Law Review, 38*, 343-367.

Lord, F., Tuler, S., Webler, T. & Dow, K. (2012). Unnecessarily neglected in planning: Illustration of a practical approach to identify human dimension impacts of marine oil spills. *Journal of Environmental Assessment Policy and Management, 14*(2), 1-23.

Plater, Z. J. B. (2011). The Exxon Valdez resurfaces in the Gulf of Mexico...and the hazards of "megasystem centripetal di-polarity". *Boston College Environmental Affairs Law Review, 38*, 391-416.

Schwartz, B. S., Parker, C. L., Hess, J. & Frumkin, H. (2011). Public health and medicine in an age of energy scarcity: The case of petroleum. *American Journal of Public Health, 101*(9), 1560-1567.

Steinzor, R. (2011). Lessons from the North Sea: Should "safety cases" come to America? *Boston College Environmental Affairs Law Review, 38*, 417-444.

Warren. C. (2014). Trouble in the melting Arctic: The EPA's failure to impose air pollution control measures. *Environmental Affairs, 41*, 118-132.

Watts, M. J. (2005). Righteous oil? Human rights, the oil complex, and corporate social responsibility. *Annual Review of Environment & Resources, 30*, 373-407.

Chapter 21

FROM ENERGY CRISIS TO FREE ENERGY

Overview

Fossil fuels such as coal, natural and shale gas and pertinent to fossil fuel factors such as application, prevalence, health and environmental hazards are elaborated upon. The process of fracking is described alongside related to fracking policies, adverse environmental consequences and public opinion. Renewable energy sources including bioenergy, wind power and the Bedini free energy generator are introduced. "From Energy Crisis to Free Energy" concludes with a comparison between light-duty vehicles and electric vehicles relative to their environmental impact.

21.1 Electricity

"Our modern society demands massive amounts of electricity every day" (Moren, 2009). Air emitted from electricity generators has a large impact on atmospheric pollution (Burtraw et al., 2005). Most electricity in the United States (U.S.) is generated from the combustion of fossil fuels such as oil, coal and natural gas (Moren, 2009).

21.1.1 Fossil Fuel

The demand for fossil fuels has increased worldwide (Fisk, 2013). The governments in the world spend roughly half a trillion U.S. dollars each year on subsidizing fossil fuel. Fossil fuels are viewed as the major cause of climate change. The subsidy of fossil fuels has hence been described as "a reckless use of public funds" (Yates, 2014).

More robust environmental protection laws in addition

333

to the limited amounts of fossil fuels have contributed to industries and scientists searching for alternative fuel types (Maina, 2014).

21.1.1.1 Coal

Coal is the least expensive, most plentiful and commonly used as an energy source fossil fuel in the world (Suriyawong et al., 2009). Coal is easily accessible, yet is naturally radioactive. The U.S. is the most abundant in coal reserves country in the world. One-fourth of all known coal worldwide is located in the U.S. (Farahani, 2014). About 60% of electrical energy generated in the U.S. is derived from the combustion of coal (Suriyawong et al., 2009). Coal is the leading energy supply in South Africa, yet alternative energy supply methods are being explored in order to reduce the greenhouse gas (GHG) emissions that result from the use of coal (Cohen & Winkler, 2014).

The demand for electricity in China has increased drastically in recent years. In 2001, 95% of the generation of thermal power in China was derived from coal plants which China has large stocks of. Coal power plants release air pollutants. Yet, "coal power plants have fallen behind in terms of adopting countermeasures to decrease pollutants." Only 12.6% of coal power generation plants have installed desulfurization technology (Ohdoko et al., 2013).

The least expensive method of recovering energy from industrial and municipality by-products is co-combustion of such products with coal. Such combustion process contributes to the emission of contaminants, including trace elements and fine particles (Suriyawong et al., 2009). According to 2008 reports by the WHO, pollution from coal particulates contributes to shortened lifespan of 1,000,000 people per year in the world of which 24,000 people live in the U.S. (Farahani, 2014). Auto and truck mechanics are exposed to coal-based products and aromatic hydrocarbons, which have been cited as risk factors for bladder cancer (Smith et al., 1985).

21.1.1.2 Natural Gas

The U.S. gas industry has created 3 million jobs and amounts to U.S. $385 billion in "direct economic activity" (Howarth & Engelder, 2011). The consumption of natural gas in the U.S. has increased exponentially in the past 30 years. Natural gas extraction in the U.S. has increased by 25% since 2005 (Hanna, 2014). Natural gas is used to power or heat 50% of U.S. households which implies a well-established infrastructure (Fisk, 2013). The consumption of natural gas in the U.S. is projected to continue increasing through 2040. Hundreds of thousands of miles of natural gas pipelines are already being used in the U.S. while thousands of addition natural gas pipelines are either in the construction or approval phase. The Federal Energy Regulatory Commission (FERC) is responsible for the review and approval of natural gas pipeline projects processes (Hanna, 2014). U.S. state governments receive millions of dollars from severance taxes on natural gas (Fisk, 2013). Currently, 50% of natural gas produced in the U.S. is derived from fracking (Howarth & Engelder, 2011).

21.1.1.2.1 *Center for Biological Diversity vs. U.S. Bureau of Land Management*

In 2009, Ruby Pipeline, L.L.C. intended to build a 678-mile long natural gas pipeline from Wyoming to Oregon. The pipeline was to pass through an endangered species habitat, threatening 9 endangered fish species. The Federal Energy Regulatory Commission and the Fish and Wildlife Service approved the proposal and the accompanying voluntary precautions designed to ensure the welfare of the 9 endangered fish species. Environmental advocate groups objected to the voluntary nature of the endangered fish species protection plan. In *Center for Biological Diversity vs.*

U.S. Bureau of Land Management, the U.S. Court of Appeals for the Ninth Circuit found that federal agencies "violated the Endangered Species Act" which requires mandatory and not voluntary conservation plans. The Endangered Species Act (ESA) mandates federal agencies to "conserve endangered and threatened species and their habitat" (Hanna, 2014).

21.1.1.3 Shale Gas

Shale gas is a proposed alternative to coal as it is characterized by lower GHG emissions (Cohen & Winkler, 2014). Shale gas consists primarily of methane. (Cohen & Winkler, 2014): Shale gas is derived from hydraulic fracturing also referred to as "fracking" (Jones et al., 2013).

21.1.1.3.1 Fracking

Fracking originates in the U.S. During fracturing, a borehole is drilled in the ground. A mixture of chemicals and water are inserted into the shale to open fractures. The shale gas flows from the paths created by the fractures to the borehole and up to the surface. The process of hydraulic fracturing was initially demonstrated in the U.S. in the 1970s. It spread commercially during the 21st century. By 2012, 40% of total natural gas production in the U.S. was attributed to shale gas (Jones et al., 2013). U.S. energy companies have used fracking since the 1940s (Davis & Hoffer, 2012). The global estimate of shale gas resources amounts to 2066 trillion cubic meters with Argentina, China, the U.S., Canada and Algeria accounting for 53% of this total. China owns the largest shale gas reserves worldwide (Jones et al., 2013).

21.1.1.3.1.1 Chemicals

80 to 300 tons of chemicals may be used for each fracking. The type of chemicals may differ from well site to well site.

U.S. regulation, particularly the 2005 Energy Act, does not mandate disclosure of chemicals being used by gas companies during fracking. Such disclosure is voluntary (Jacobson, 2012).

21.1.1.3.1.2 Pollution

Industrial companies state that they have used fracking to establish over 1 million natural gas and oil wells in the world since the late 1940s. Natural gas is being promoted as a clean energy source, yet "the greenhouse-gas footprint of shale gas is worse than that for coal or oil." Between 1.7% and 7.9% of methane is leaked into the atmosphere from the shale gas well head, storage facility or pipeline as part of the total production of a shale gas well. Concerns have been raised about the effects of leaked into the atmosphere methane on global warming. Methane has a brief half-life in the atmosphere. Carbon dioxide is further emitted into the atmosphere from the shale gas development and from the combustion of shale gas for heating.

Further, contamination associated with fracking processes could result from blowouts, improper fracking fluids disposal and surface spills. Two major environmental concerns associated with fracking include water use and pollution. "Millions of gallons of water are required to stimulate a well." Potentially harmful to the environment materials contained in shale gas such as radioactive isotopes and salts of barium are brought through the well pipe to the surface via flowback water. Even though such waters are presently treated prior to disposal, there is a risk that they will leak into water streams or groundwater upon reaching the surface untreated as a result of blowouts, leaking well heads or holding tanks. Methane could further indirectly leak into groundwater (Howarth & Engelder, 2011). Shale gas leakages are a possibility as each shale gas extraction site has between 55 and 150 pipe connections (Cohen & Winkler, 2014).

Opponents to fracking and environmentalists propose that fracking contributes to the emission of substantial quantities of methane (a GHG) as well as the contamination of surface and ground waters. A study demonstrated the presence of chemicals such as radium and barium in fracked water at concentrations higher than drinking water standards. Improper disposal of fracked water could contribute to the release of fracked water chemicals into drinking water while posing a risk to aquatic life. Another study established a link between fracking and compromised water quality (Fisk, 2013).

Information pertaining to health threats and environmental risks induced by fracking such as drinking water contamination as fracking fluid leaks into water systems and soil pollution caused by water spill from fracking waste are growing. One study identified 632 chemicals used during hydraulic fracturing. 75% of identified chemicals could have an effect on the skin, eyes, other sensory organs, gastrointestinal and respiratory systems. A number of these chemicals could also affect the nervous, brain, immune, cardiovascular, endocrine and respiratory systems and exert mutagenic and carcinogenic effects (Jacobson, 2012). A different study found that between 2005 and 2009, several of the 14 largest gas and oil companies in the U.S. used 29 potentially cancerogenic chemicals in their fracking products. Some of these companies used "extremely toxic" chemicals, including lead and benzene (White, 2014).

Experts disagree on the actual risks of fracking while "the public holds both a conflicted but ultimately limited view of fracking." It is likely that fracking companies intentionally hide pertaining information as a means to prevent experts from arriving at a consensus regarding the objective risks of fracking. It is questionable as to whether the lack of consensus has hindered the demand for the disclosure of information. It is likely that a large proportion of the public has not formulated an opinion regarding the risks of fracking due to the lack of such consensus (Fisk, 2013).

A 2010 national survey revealed that 45% of Americans were aware that fracking was a controversial topic while 69% expressed concerns about the effects of fracking on drinking water safety. The majority of respondents would support disclosure requirements and additional research pertaining to the effects of fracking on health and the environment. It was recently found out that Americans welcome the economic benefits related to fracking, yet are concerned about the effects of fracking on the environment (Davis & Hoffer, 2012).

21.1.1.3.1.3 The UK

"The UK government and the business community have been keen to publicly promote the economic benefits that the development of shale gas could generate." The creation of employment opportunities, reduction of imports, tax revenue generation and support for the British manufacturing were proposed by the Institute of Directors in May, 2013 as economic benefits of United Kingdom (UK) shale gas exploitation.

There are increasing concerns regarding the environmental and social impact associated with the exploitation of shale gas via hydraulic fracturing or fracking. Opponents of shale gas exploitation argue that shale gas burning could contribute to "catastrophic climate change." Environmental concerns associated with hydraulic fracturing include groundwater and surface water pollution, air pollution and resulting health risks, "the danger of earth tremors and land subsidence." The UK government has attempted to counteract such concerns by stating its intent to encourage a safe shale gas industry that does not pose environmental risks. Opponents to fracking argue that regulation could make hydraulic fracturing safer, yet it would not be able to make it safe. Public opposing to fracking is generally perceived as an impediment to the commercial exploitation of shale gas resources via fracking. KPMG proposed that "the industry needs to control reputation risk and turn public opinion round" (Jones et al., 2013).

21.1.1.3.1.4 Fracking Violations

1056 fracking-related environmental violations and 669 traffic citations were identified between 2008 and 2010 in the Marcellus Shale region in Pennsylvania. Some of the violations included improper discharge of waste materials, construction and waste management in addition to violations of local environmental laws. Such violations contribute to increased risk of groundwater and air pollution (White, 2014).

21.1.1.3.1.5 Fracking Regulations

"Lenient" fracking-related federal and state regulations further contribute to increased risks associated with fracking. The Safe Drinking Water Act, EPA's "central authority for protecting drinking water" exempts fracking from the Underground Injection Control program, which is designed to regulate the "subsurface emplacement of fluid." A number of federal environmental acts authorize states to establish fracking regulations. While some states have set in place stricter regulations pertaining to fracking, others have failed to do so (White, 2014).

     "Shale gas competes for investment with green energy technologies, slowing their development and distracting politicians and the public from developing a long-term sustainable energy policy" (Howarth & Engelder, 2011).

21.1.2 Renewable Energy

Biofuels and wind power are considered to be renewable energy sources (Mugido et al., 2014). In the last ten years, the

U.S. and European Union (EU) have established biofuel policies designed to reduce reliance on fossil fuels and offset global warming (Gohin, 2014).

21.1.2.1 Biofuels

Biofuels, generally referred to as bioenergy, have been promoted as an environmentally sound renewable energy source that decreases the reliance on fossil fuels. Bioenergy is defined as a renewable form of energy derived from biological sources such as wood and agricultural sources that can be used to generate electricity, heat and transportation fuel. Soy, palm oil, corn or sugar cane are "the commercially established and most abundant sources of biofuel today" (Mugido et al., 2014). At present, the primary biofuel in the world is corn-based ethanol. U.S. corn production has increased by 16.38% and global corn production – by 4.32% (Gohin, 2014). Bioethanol and biodiesel are the most widely used renewable biofuels in the world. Bioethanol is used in ignition engines and biodiesel - in diesel engines (Bunger et al., 2012).

Biodiesel, produced from renewable lipid feedstock such as animal fat or vegetable oil (Maina, 2014), including palm, canola, sunflower, coconut and peanut oil (Bunger et al., 2012) is one proposed to fossil fuels alternative. Biodiesel is non-toxic, renewable, sulphur-free and biodegradable. It can be used in diesel engines and contributes to a reduction in global warming gas emissions (Maina, 2014). Diesel engine emissions (DEE) consist of nanoscale particles that are easily respirable and contain hundreds of chemicals with possible mutagenic and carcinogenic properties, including PAHs. Studies have demonstrated a causal link between DEE and acute health outcomes, particularly the worsening of asthma. A causal relationship between DEE and chronic health effects (i.e., childhood asthma; compromised lung function; cardiovascular mortality and morbidity) has also been established. Chronic inhalation of DEE could contribute to an increased risk for lung cancer and lung damage (Bunger et al., 2012).

Ethical concerns have been expressed regarding bioenergy development (Mugido et al., 2014). Biofuels "may underperform" if they contribute to alterations in land use that result in additional GHG emissions (Gohin, 2014). Bioenergy development may contribute to deforestation and other land use alterations that lead to emission of GHGs. Land has been identified as a scarce resource. Bioenergy production has been criticized for competing with food production in terms of land use as it displaces agricultural crops. Land allocated to bioenergy production could contribute to increased risks of food insecurity particularly in relevance to people living in poverty. Land use is the primary challenge to bioenergy production.

Jean Zaigler, the United Nations (UN) Special Reporter on the Right to Food (2000-2008) stated in 2007: "It is a crime against humanity to divert arable land to the production of crops which are then burned for fuel." 450 pounds of corn are required to fill a 25-gallon SUV tank with pure ethanol, equal to providing one individual with sufficient calories for an entire year (Mugido et al., 2014).

21.1.2.2 Wind Power

To meet Kyoto obligations of focusing on cleaner renewable energy sources as an alternative to traditional energy sources, the British Government has expanded its efforts in advancing the wind energy sector which, according to literature, is more beneficial in comparison with alternative renewable energy methods (Strachan et al., 2006). Wind power is promoted as an energy source that does not contribute to substantial GHG emissions, air or water pollution, or radioactivity.

After natural gas, "wind is the second fastest energy source in the United States" (Holzman, 2007). The wind energy sector has created 95,000 to 100,000 employment opportunities globally. The global wind energy market turnover in 2002 and 2003 was estimated at close to 7 billion Euros. The top ten wind companies contributed to 95% of the new wind installations in 2003. The capacity of generated wind power has increased four times in the past five years. While wind energy is considered to provide an "inexhaustible supply of clean electricity," 1 billion pounds would be required to maintain traditional back-up power supply in the UK each year (Strachan et al., 2006).

21.1.2.3 Free Energy

The Bedini free energy generator is a free energy system composed of a compact DC motor powered by a 12 VDC battery that does not run down. Researchers have duplicated the construction and demonstrations of the Bedini free energy generator (Bedini, 1991).

21.2 Vehicles

About 22 million vehicles are manufactures in the U.S. every year (Suriyawong et al., 2009). Over 2 billion new cars are expected to be in operation by 2050 (Berteaux, 2013). The ownership of light-duty vehicles is projected to increase from 700 million to 2 billion between 2000 and 2050.

Light-duty vehicles contribute to about 10% of GHG emissions and energy use worldwide. The EU and the U.S. have promoted the introduction of electric vehicles (EVs). A consortium of the International Energy Agency (IEA) and 8 countries, including Germany, China, Japan, South Africa, Spain, Sweden, France and the U.S. have established an objective to manufacture 20 million hybrid EVs by 2020. There is a general view that EVs are more environmentally friendly compared to conventional fuel-based vehicles. In truth, the production of EVs contributes to substantial increases in human toxicity, freshwater toxicity and eutrophication as well as to metal depletion. The production phase of EVs is therefore more environmentally taxing compared to internal combustion engine vehicles (ICEVs). Terrestrial acidification potential (TAP) is comparable between ICEVs and EVs production. As part of ICEVs, the global warming potential (GWP) impact occurs primarily during the fuel combustion phase while the majority of GWP impact of EVs takes place during the production phase. GWP impact from the production of EVs is approximately twice the GWP of ICEVs. GWP from EV production is close to twice the GWP impact potential reported by past studies. EVs provide the ground for moving GHG emissions "away from the road rather than reducing them globally." Only the use of clean energy could contribute to environmental benefits relative to EVs. Alternative transportation technologies need to also be taken into considered to ensure long-term benefits (Hawkins et al., 2012).

Discussion

"Paradoxical Formation" suggested that global warming and climate change are taking place at unprecedented rates and are projected to continue to accelerate. The main or root problem is the combustion of fossil fuel leading to GHG emissions and subsequent global warming and climate change. Yet, only an abstract and not the full article has been made available for the Bedini free energy generator which appears to be a promising alternative to fossil fuels. There is an explanation for such anomaly. According to Sosé's theory, free energy would signify zero profit and power for industries and governments which is in exact opposition to their agenda.

Furthermore, while EVs are advertised as environmentally friendly (i.e., deceit and a problem), in reality they are as detrimental to the environment as traditional vehicles are. It could be deduced that such a deceit is manufactured in order for humanity to purchase electrical vehicles and generate profit for the industry. At the same time, humanity would be unknowingly creating environmental problems which would subsequently increase governmental and industrial profit in a spectrum of ways touched upon earlier in "Sovereign Terra."

Similarly, biofuels are advertised as environmentally friendly (i.e., deceit and a problem), yet the truth is exactly the opposite. Profit is not only generated from selling biofuels, but similarly to EVs, it could be deduced that profit is also being generated from the environmental and human health problems that are created through their use.

The biofuel and EVs examples resemble the legal hunting deceit reviewed under "Ghostly Habitat." An illusion is created that hunting is necessary for animal population control. The public is the first to respond to such manipulation. People are hunting while governments and industries are collecting the profit from hunting activities and subsequent environmental imbalances. Similarly, people are the ones polluting the environment via the use of EVs and biofuels while profit is being collected by governments and industries as a result of their use. Unknowingly having been trapped by the deceit that biofuels and EVs are environmentally friendly, humanity itself is inflicting severe environmental damages.

Similarly to nano technology, GMOs, food aid, geoengineering, meat and packaged salad greens production, truth is hidden from the public regarding chemicals used in fracking. Such secrecy not only suggests that poisonous chemicals are used in fracking, but the deceit also ensures that humanity would continue using fracking-derived products, generating profit for the industry and governments.

"Conditions in Motion" lastly stated that earthquakes result from fracking, which creates yet another problem and additional opportunities for profit (i.e., fundraising).

References

Bedini, J. C. (1991). Bedini Free energy generator. *Proceedings of the Intersociety Energy Conversion Engineering Conference, 4*, 451-456.

Berteaux, D. (2013). Quebec's large-scale Plan Nord. *Conservaion Biology, 27*(2), 242-247.

Bunger, J., Krahl, J., Schroder, O., Schmidt, L. & Westphal, G. A. (2012). Potential hazards associated with combustion of bio-derived versus petroleum-derived diesel fuel. *Critical Reviews in Toxicology, 42*(9), 732-750.

Burtraw, D., Evans, D. A., Krupnick, A., Palmer, K. & Toth, R. (2005). Economics of pollution trading for SO2 and NOX. *Annual Review of Environment & Resources, 30*, 253-289.

Cohen, B. & Winkler, H. (2014). Greenhouse gas emissions from shale gas and coal for electricity generation in South Africa. *South African Journal of Science, 110*(3/4), 1-5.

Farahani, J. V. (2014). Man-made major hazards like earthquake or explosion; Case study Turkish mine explosion. *Iranian Journal of Public Health, 43*(10), 1444-1450.

Fisk, J. M. (2013). The right to know? State politics of fracking disclosure. *Review of Policy Research, 30*(4), 345-365.

Hanna, B. (2014). The Ninth Circuit constrains non-enforceable public-private endangered species conservation agreements. *Environmental Affairs, 41*, 42-53.

Hawkins, T. R., Singh, B., Majeau-Bettez, G., Stromman,A. H. (2012). Comparative environmental life cycle assessment of conventional and electric vehicles. *Journal of Industrial Ecology, 17*(1), 53-64.

Holzman, D. C. (2007). The vanadium advantage. *Environmental Health Perspective, 115*(7), A359-A361.

Howarth, R. W. & Engelder, T. (2011). Should fracking stop? *Nature, 477*, 271-275.

Jones, P., Hillier, D. & Comfort, D. (2013). Fracking and public relations: rehearsing the arguments and making the case. *Journal of Public Affairs, 13*(4), p. 384-390.

Maina, P. (2014). Engine emissions and combustion analysis of biodiesel from East African countries. *South African Journal of Science, 110*(3/4), 1-8.

Moren, H. (2009). The difficulty of fencing in interstate emissions: EPA's Clean Air Interstate Rule fails to make good neighbors. *Ecology Law Quarterly, 36*, 525-552.

Mugido, W., Blignaut, J., Joubert, M., De Wet, J. D., Knipe, A., Joubert, S., Cobbing, B., Jansen, J., Le Maitre, D. & Van Der Vyfer, M. (2014). Determining the feasibility of harvesting invasive alien plant species for energy. *South African Journal of Science, 110*(11/12), 45- 50.

Ohdoko, T., Komatsu, S. & Kaneko, S. (2013). Residential preferences for stable electricity supply and a reduction in air pollution risk: A benefit transfer study using choice modeling in China. *Environmental Economics & Policy Studies, 15*, 309-328.

Smith, E. M., Miller, E. R., Woolson, R. F. & Brown, C. K. (1985). Bladder cancer risk among auto and truck mechanics and chemically related occupations. *American Journal of Public Health, 75*(8), 881-883.

Strachan, P. A., Lala, D. & Malmborg, F. V. (2006). The evolving UK wind energy industry: Critical policy and management aspects of the emerging research agenda. *European Environment, 16*, 1-18.

Suriyawong, A., Magee, R., Peebles, K. & Biswas, P. (2009). Energy recycling by co-combustion of coal and recovered paint solids from automobile paint operations. *Journal of the Air & Waste Management Association, 59*, 560-567.

White, J. M. (2014). Whiteman v. Chesapeake: Damage to human health and the environment as seen through an application to hydraulic fracturing. *Ecology Law Quarterly, 41*(645), 645-652.

Yates, R. (2014). Recycling fuel subsidies as health subsidies. *Bull World Health Organization*, *92*, 547A.

Chapter 22

DNA MUTATION

Overview

Sources of radio-frequency radiation (RFR), ionizing radiation and microwave radiation are introduced alongside adverse health effects resulting from exposure to such sources of radiation.

***

Humanity today is exposed to RFR emitted from sources such as microwave ovens, television and radio transmitters, mobile phones (MPs) and mobile phone stations (Ozgur et al., 2010).

22.1 Mobile Phone Radiation

At present, MPs are substituting land-line phones "and becoming ubiquitous in our lives" (Colak et al., 2012). "Mobile phones form an integral part of the daily life of billions of users worldwide and this number is ever-increasing" (Stander et al., 2011). Younger people are increasingly using mobile telephony (Wood et al., 2006). Concerns have been expressed regarding the long-term health effects of MP use (Stander et al., 2011), particularly the association between mobile phone use, cancer and the brain (Anin et al., 2004).

A number of recent studies reveal a significant increase in DNA damage in cells following RFR exposure from MP use. "Over accumulation of genetic damage in cells in time can be the triggering fact of cancer and cell death." "Continuous long-term exposure to RFR is rapidly increasing in the environment" (Cam & Seyhan, 2012). A review of epidemiological studies revealed a link between mobile telephone use longer than 10 years and increased risk of acoustic neurinomas and ipsilateral gliomas (Markova et al., 2010).

Third-generation (3G) MPs are part of the world's largest mobile telecommunication systems. 3G MP applications such as messaging, video calling and conferencing emit steadier electromagnetic radiation (EMR) into the environment. Studies reveal that MPs EMR may contribute to a number of health conditions including headache, damage to the inner ear, blurred vision, inflammation of the eyes, retina oxidative stress, great irritation, increase in forgetfulness and carelessness, heart rate (HR) fluctuations as a result of occupational exposure to EMR and possible impact on the circulatory system. It has also been hypothesized that very low-frequency (40-50 Hz) EMR could suppress the production of melatonin from the pineal gland (Colak et al., 2012).

22.2 Radiotherapy & Chemotherapy

"Approximately 42% of the U.S. population die from cancer-related causes" (Martin & Semelka, 2007). Studies have revealed a link between radiotherapy in combination with chemotherapy and cancer. A number of studies have also demonstrated an increased risk of secondary acute leukemia in patients treated with combined radiotherapy and chemotherapy compared with a treatment of radiotherapy alone. The Environmental Protection Agency (EPA) has established general guidelines for assessing the risk of chemical mixtures. Yet, exposure to the combination of different agent classes (i.e., radiation and chemical) "has not yet been addressed in any substantive way" (Chen et al., 2001).

22.3 Ionizing Radiation

Medical diagnostic X-ray machines generate ionizing radiation

(Martin & Semelka, 2007). "Ionizing radiation is defined as radiation that has enough energy to displace electrons from atoms inside of the patient" (Moore, 2014). Ionizing radiation has been listed as a known human carcinogen as part of the 2005 report from the Department of Health and Human Services. Ionizing radiation alters and damages cellular DNA which, in turn, induces neoplasms or tumors. Cancer risk from X-ray radiation is equivalent to cancer risk from atomic bomb radiation for an equivalent dose. Leukemias have been the most widely studied tumors in relation to human exposure to ionizing radiation (Martin & Semelka, 2007). The greater the exposure to radiation, the greater the likelihood that the individual will develop cancer (Moore, 2014). Even the smallest dose of low-level ionizing radiation could increase the health risks to human beings (Martin & Semelka, 2007).

A 2009 report by the National Council on Radiation Protection & Measurements revealed that the exposure to radiation, particularly from medical procedures, among the United States (U.S.) population has doubled in the last two decades. 67 million computed tomography (CT) scans, 18 million nuclear medical and 17 million fluoroscopy procedures were performed in the U.S. in 2006. These figures are projected to continue to increase. "Although radiation procedures often are necessary, they become controversial when ordered in abundance" (Moore, 2014).

Over 60 million CT examinations, equivalent to 70% of total medical X-ray exposure, are performed in the U.S. each year. Lifetime attributable risk (LAR) for developing cancer or leukemia as a result of a single dose of 10 mSv is 1 in 1000. Given that 60 million CT examinations are performed each year, 60,000 leukemia and solid cancer cases are estimated to develop annually as a result of 1 whole-body 10 mSv examination dose per year. Exposure to ionizing radiation doses over 100 mSv has been associated with the development of non-neoplastic conditions such as cardiovascular disease. 600,000 abdominal (i.e., 10 mSv) and head CT examinations are administered in children under 15 years of age each year, resulting in estimated LAR of 500 deaths from cancer.

The majority of institutions require patients to sign an informed consent prior to undergoing a CT examination. Yet, a discussion of radiation risks with the patient is not mandatory and is rarely included as a component of the consent form. The informed consent further requires the radiologist to elaborate on CT alternatives, yet such procedure is also not commonly undertaken (Martin & Semelka, 2007). One study revealed that 34% of patients were not aware that CT scans expose their bodies to radiation (Moore, 2014).

22.4 Microwave Radiation

Microwave ovens have become widespread in developed countries and to a certain extent, in developing nations. Microwave ovens are commonly used to cook or warm foods (Park et al., 2006).

A study investigated the residual content and transfer of bisphenol A from 18 brands of polycarbonate baby bottles after 3 microwave heating cycles. Findings demonstrated that residual content ranged from 1.4 to 35.3 mg kg^{-1} while migration was estimated to be in the range of <0.1 to 0.7 µg l^{-1}. Migration levels were below bisphenol A migration limits set forth by the Commission Directive 2004/19/EC (Ehlert et al., 2008).

22.5 Atomic Bomb Radiation

Results from the present study revealed that atomic bomb radiation is associated with aortic atherosclerosis (Yamada et al., 2005).

"Malignant resources are exhausting us."

By Sosé Gjelaj

Discussion

Similarly to nano technology, GMOs, fracking, etc., it could be concluded that the truth about MPs, microwave and medical technology radiation is not revealed to humanity as it will prevent public utilization of harmful devices and therefore the generation of profit for governments and industries. The problems generated by radiation (i.e., cancer) create additional opportunities for profit (i.e., pharmaceutical and health care sector).

Also and in similarity to industrial and pharmaceutical contaminants, radiation is toxic. It is difficult to fathom that technology has evolved so rapidly in the past decades and yet environmentally friendly products have not been introduced. The only reasonable explanation for such irony is the actual intent of governments and industries to refrain from utilizing beneficial to human and environmental health products as health and well-being signifies no opportunities for profit and power.

References

Anin, J.-M., Carrere, N., Chalan, Y., Dulou, P.-E., Larrieu, S., Letenneur, L., Veyret, B. & Dulon, D. (2004). Effects of exposure of the ear to GSM microwaves: In vivo and in vitro experimental studies. *International Journal of Audiology, 43*, 545-554.

Cam, S. T. & Seyhan, N. (2012). Single-strand DNA breaks in human hair root cells exposed to mobile phone radiation. *International Journal of Radiation Biology, 88*(5), 420-424.

Chen, W.-c. & McKone, T. E. (2001). Chronic health risks from aggregate exposures to ionizing radiation and chemicals: Scientific basis for an assessment framework. *Risk Analysis, 21*(1), 25-42.

Colak, C., Parlakpinar, H., Ermis, N., Tagluk, M. E., Colak, C., Sarihan, E., Dilek, O. F., Turan, B., Bakir, S. & Acet, A. (2012). Effects of electromagnetic radiation from 3G mobile phone on heart rate, blood pressure and ECG parameters in rats. *Toxicology and Industrial Health, 28*(7), 629-638.

Ehlert, K. A., Beumer, C. W. E. & Groot, M. C. E. (2008). Migration of bisphenol A into water from polycarbonate baby bottles during microwave heating. *Food Additives and Contaminants, 25*(7), 904-910.

Markova, E., Malmgren, L. O. G. & Belyaev, I. Y. (2010). Microwaves from mobile phones inhibit 53BP1 focus formation in human stem cells more strongly than in differentiated cells: Possible mechanistic link to cancer risk. *Environmental Health Perspectives, 118*(3), 394- 399.

Martin, D. R. & Semelka, R. C. (2007). Health effects of ionizing radiation from diagnostic CT imaging: Consideration of alternative imaging strategies. *Applied Radiology, 36*(5), 20-29.

Moore, Q. T. (2014). Medical radiation dose perception and its effect on public health. *Radiologic Technology, 85*(3), 247-255.

Ozgur, E., Guler, G. & Seyhan, N. (2010). Mobile phone radiation-induced free radical damage in the liver is inhibited by the antioxidants n-acetyl cysteine and epigallocatechin-gallate. *International Journal of Radiation Biology, 86*(11), 935-945.

Park, D-K., Bitton, G. & Melker, R. (2006). Microbial inactivation by microwave radiation in the home environment. *Journal of Environmental Health, 69*(5), 17-24.

Stander, B. A., Marais, S., Huyser, C., Fourine, Z., Leszczynski, D. & Joubert, A. M. (2011). Effects of non-thermal mobile phone radiation on breast adenocarcinoma cells. *African Journal of Science, 107*(9/10), 47-55.

Wood, A. W., Loughran, S. & Stough, C. (2006). Does evening exposure to mobile phone radiation affect subsequent melatonin production? *International Journal of Radiation Biology*, *82*(2), 69-76.

Yamada, M., Naito, K., Kasagi, F., Masunari, N. & Suzuki, G. (2005). Prevalence of atherosclerosis in relation to atomic bomb radiation exposure: An RERF Adult Health Study. *International Journal of Radiation Biology*, *81*(11), 821-826.

Chapter 23

WASTE TEMPERING

Overview

"Waste Tempering" elaborates on e-waste management and recycling, including metal recycling.

23.1 Waste Management

Waste management contributes to 2% of worldwide emissions of greenhouse gas (GHG) responsible for climate change. Methane emissions from landfills and carbon dioxide discharges from waste combustion are the primary GHG released during waste management. The discussion regarding optimal waste management options has been ongoing. "Contradicting claims have frequently been in focus in the waste management debate" (Tyskeng et al., 2010).

23.1.1 E-Waste

Electronic and electric waste, also referred to as e-waste, is a major global environmental health problem due to its enormous production and inadequate management policy. Improper e-waste management has contributed to the pollution of water, soil and air with toxic substances such as organic compounds and heavy metals (Tang et al., 2013).

It has been estimated that about 500 million computers became "obsolete" in the United States (U.S.) between 1997 and 2007. 80% of U.S. e-waste is re-directed to Asia and Africa. The U.S. is the only developed country which has not taken part in the United Nations (UN) Basel Convention, banning the export of hazardous waste material to developing nations. Alongside New Delhi in India, Guiyu in

China is one of the popular e-waste settings.

Each year, Guiyu processes millions of tons of domestic and international e-waste. Due to the high expense of safe and clean recycling techniques, the recycling processes and activities in Guiyu are "very primitive," resulting in the discharge of e-waste residues in ponds, rivers, riverbanks, open fields, roadsides, irrigation canals, yards and workshops. The release of hazardous substances as part of e-waste management could have adverse health consequences. Lead (Pb) is the most commonly used poisonous heavy metal in electronic devices and is considered as one of the main heavy metal pollutants during e-waste recycling. It infiltrates biological systems via soil, air, water and food. Children are especially prone to Pb poisoning as they absorb greater Pb quantities compared to adults.

The blood lead levels (BLLs) of children living in Guiyu, China, and those living in the town next to Guiyu, Chendian, also located in China, were compared. Findings revealed that the BLLs of children living in Guiyu ranged between 4.40 µg/dL and 32.67 µg /dL with a mean of 15.3 µg/dL. 81.8% of children had BLLs greater than 10 µg/dL. BLLs of children who lived in Chendian ranged between 4.09 µg/dL and 23.10 µg/dL with a mean of 9.93 µg/dL. Children living in Guiyu had significantly higher BLLs compared to those living in Chendian. The higher BLLs of children who lived in Chendian further suggested that Pb contamination from Guiyu could have been transported to Chendian via air, dust and water. Pb contamination from the processing of e-waste seems to have reached BLLs, considered to be "a serious threat to children's health around the e-waste recycling area" (Huo et al., 2007).

23.1.2 Asia

As economic growth is rapidly taking place in Asia, generation and management of solid waste "is becoming a major social and environmental issue." The urban areas of Asia are

357

estimated to generate approximately 760,000 tons of municipal solid waste (MSW) daily. This figure is anticipated to increase to 1.8 million tons of waste by 2025. Due to its inexpensive nature, landfilling is the primary method of waste disposal in a number of Asian countries. In general terms, developed countries generate higher levels of MSW per unit (Terazono et al., 2005).

23.1.3 Recycling

Recycling is considered as an overall environmentally preferable waste management method in comparison to landfill or incineration waste processes.

The overall academic literature on pro-environmental conduct and the reasons why people do or do not recycle is widespread, yet scattered in a variety of journals while "results are frequently contradictory." Findings from the literature reveal that the values important to people who recycle include environmental responsibility, self-actualization, collectivist rather individualist orientation, respect, helpfulness, accomplishment and achievement. It was also demonstrated that people who felt responsibility for the environment were more likely to take part in environmentally-sensitive activities compared to people who did not. For most people, recycling was perceived as a moral behavior. Findings from the U.S. further revealed that people do not recycle as they lack the information needed to know how to do so. It is therefore essential that people are aware of how to recycle waste. According to the theory of reasoned action (TRA), the intention to act precedes the actual behavior. In light of recycling, TRA implies that the intention to recycle determines actual recycling behavior. Social norms are behavioral standards expected by society while personal norms are a person's individual standards of behavior (Smallbone, 2005).

23.1.3.1 Metal

About 50 million metric tons (t) of steel was produced in Germany each year prior to the economic crisis of 2007 (Frohling et al., 2012). Particular metal products and their alloys are used for a number of years. When no longer in use, such products are either recycled for secondary production or discharged into the environment. Recycling and re-melting of metals is less energy consumptive compared to producing primary metal from its mineral form. In theory, metals can be recycled infinitely, maintaining their atomic properties (Johnson et al., 2013).

73% of the global aluminum produced since the 1950s is still in use in 2003. About 68% to 69% of metallic aluminum present in the U.S. since 1900 is still in use – 67% in domestic use and 1% to 2% allocated to scrap. 6% to 7% of it is "lost to the environment" while 25% of the aluminum is "unknown." It could have been exported, hibernating or lost during collection (Chen, W.-Q., 2013).

23.1.3.2 New York City

In the aftermath of 9/11, the Mayor of New York City, Michael Bloomberg, proposed a temporary suspension of glass and plastic recycling in New York City in order to close projected billions of dollars budget gap. The plan was perceived as anti-environmental by many. In reality, however, as also noted by the Mayor of New York City, the glass and plastics recycling program in New York City was not efficient. Almost 90% of the plastic and glass "set aside by thoughtful New Yorkers" was transported to landfills and not to recycling facilities as intended. Such occurrence is not unique to New York City. The U.S. Environmental Protection Agency (EPA) states that only 5.4% of plastic produced in the U.S. was recycled in 2000. "It is an environmental paradox that the United States is digging up new oil fields and coal mines in pristine areas and, at the same time, converts greenfields to brownfields by burying nearly 20 million tons of fuel in the form of plastics" (Themelis & Todd, 2004).

Discussion

Deriving from Sosé's theory, e-waste recycling in Guiyu, China, represents another example of how problems are created in the name of profit. The lack of efficient recycling equipment prevents proper e-waste recycling. The result is the pollution of the environment which, as was made apparent earlier in "Sovereign Terra," leads to disease and thus profit for governmental and industrial sectors.

A different problem was noted as part of the New York City recycling example. Not only is recyclable material not being recycled (i.e., problem) which translates into direct profit without the provision of service, but material is being buried underground, contributing to the pollution of the environment (i.e., problem). It is apparent that alternative motives (i.e., pollution of the environment) leading to profit are prioritized. Means to profit from a polluted environment are similar to the ones described earlier in "Sovereign Terra" (i.e., pharmaceutical and health care sectors). The created problem is also apparent in the fact that oil drilling is taking place at an unprecedented rate, yet potential energy from recycling material is not being used to produce energy; rather, it is buried under the ground.

Lastly, the location of 25% of aluminum is unknown. The truth is yet again hidden from humanity. From the evidence provided thus far, it could be concluded that profit is being generated from such secrecy. The source of profit cannot be deduced from reviewed research.

References

Chen, W.-Q. (2013). Recycling rates of aluminum in the United States. *Journal of Industrial Ecology, 17*(6), 926-938.

Frohling, M., Schwaderer, F., Bartusch, H. & Schultmann, F. (2012). A material flow-based approach to enhance resource efficiency in production and recycling networks. *Journal of Industrial Ecology, 17*(1), 5-19.

Huo, X., Peng, L., Xu, X., Zheng, L., Qiu, B., Qi, Z., Zhang, B., Han, D. & Piao, Z. (2007). Elevated blood lead levels of children in Guiyu, an electronic waste recycling town in China. *Environmental Health Perspectives, 115*(7), 1113-1117.

Johnson, J. X., McMillan, C. A., & Keoleian, G. A. (2013). Evaluation of life cycle assessment recycling allocation methods: The case study of aluminum. *Journal of Industrial Ecology, 17*(5), 700-711.

Smallbone, T. (2005). How can domestic households become part of the solution to England's recycling problems? *Business Strategy and the Environment, 14*, 110-122.

Tang, X., Qiao, J., Chen, C., Chen, L., Yu, C., Shen, C. & Chen, Y. (2013). Bacterial communities of Polychlorinated Biphenyls polluted soil around an E-waste recycling workshop. *Soil and Sediment Contamination, 22*, 562-573.

Terazono, A., Moriguchi, Y., Yamamoto, Y. S., Sakai, S. I., Inane, B., Yang, J., Siu, S., Shekdar, D. L., Idris, A. B., Magalang, A. A., Peralta, G. L., Lin, C. C., Vanapruk, P. & Mungcharoen, T. (2005). Waste management and recycling in Asia (2005). *International Review for Environmental Strategies, 5*(2), 477-498.

Themelis, N. J. & Todd, C. E. (2004). Recycling in a megacity. *Journal of the Air & Waste Management Association, 54*, 389-395.

Tyskeng, S. & Finnveden, G. (2010). Comparing energy use and environmental impacts of recycling and waste incineration. *Journal of Environmental Engineering, 136* (8), 744-748.

Chapter 24

DOWNWARD ACCELERATION

Overview

"Downward Acceleration" focuses on economic aspects including homelessness, food and oil price shocks, income inequality, corruption, unemployment, capitalism and United States (U.S.) immigrants.

24.1 Homelessness

In the U.S., 633,782 people are homeless every night. Over 15% of these people are chronically homeless while more than 18% of homeless people are older than 50 years. Men at the highest risk of homelessness are those who are veterans and those who are between 45 and 54 years of age. "New York City is experiencing an all-time high of homelessness." Over 28,000 adults spend every night in a shelter while about 3,000 people live on the streets. Homeless people experience high rates of mental illness, substance abuse and physical disease (Asgary et al., 2014).

24.2 Food & Oil Price Shock

An uncontrollable by authorities event that has a downslope effect on the economy and results from conditions such as climate or political change represents an external shock. Shocks in food prices and commodity prices (i.e., world oil prices), individually and together, prevent global economic growth. Food and oil prices are interdependent while the increase in both is the cause for a worldwide inflation. Rising of oil prices, for example, increases manufacturing expenses, decreases industrial production, leads to a decline in food import and export and has a significant impact on interest and exchange rates. A good example are the food and oil prices shocks in the 1970s resulting in worldwide economic crises and culminating in famine lasting from 1973 until 1974. Greater reliance on alternative renewable energy sources and local food production in addition to expanding food reserves are some of the solutions proposed for resource-poor countries such as Korea, Taiwan and Thailand which are most vulnerable to oil and food price fluctuations (Alom et al., 2013).

24.3 Income Inequality

Evidence reveals high global income inequality is "not trending down over time." The national wealth data from 20 countries was analyzed. These 20 countries are the wealthiest and largest countries in the world, encompassing 59% of the world's population and 75% of global wealth. The U.S. is the wealthiest country with purchasing power parity (PPP) U.S. $201,319 per adult in 2000. India, on the opposite side of the spectrum, has PPP U.S. $12,201 per adult. "Global wealth-holding is highly concentrated, much higher than in the case of income. The share of the top 10% of adults in 2000 is estimated to be 70.7%...The share of the bottom half is just 3.7%." The main reason for high global wealth inequality may be the high wealth inequality within countries. The bottom 30% owned 1% of global wealth altogether while the top 1% owned 32 times as much wealth as the bottom 30% (Davies et al., 2010).

24.4 Corruption

Corruption may contribute to money laundering and additional criminal activities. It compromises the principles of equality, democracy and the rule of law. The model for police career advancement "may serve as a cause for corruption" in the majority of police departments. "An environment in which many officers either accept the existence of corruption or participate in corrupt activities calls into question the quality of leadership." Nations with greater degrees of authoritarian regimes are more likely to experience higher rates of corruption. Globalization not only resulted in expanded trade opportunities, but also in greater possibilities for corrupt activities. Prosecution of cross-border corruption rendered challenges for authorities due to the disparities in the legal systems of various nations. Not only political, but economic equality as well is necessary to deter corruption (Ionescu, 2012).

24.5 Unemployment & Capitalism

On a global scale, "the power and influence of the nation-state is decreasing, under pressure from flows of capital, labour and services across international boundaries" (Franks & Cleaver, 2007).

The primary aim of economics is to "improve people's lives" and to cause "our world to move forward" (Ionescu, 2012). The global economic crisis that took place in 2008 appears to have contributed to increased rates of suicide, particularly among American and European men (Chang, 2013). Unemployment impacts both societies and individuals negatively. In 2009, global unemployment rate reached 212 million, an increase of 34 million in comparison to 2007. There were additional 4,884 suicides in 2009 compared to the expected rate based on data from previous years. Increase in suicide rates were positively correlated with an increase in unemployment rates, especially among nations with already low unemployment rates prior to the crisis of 2008. Capitalism excludes millions of people from its system at any given time while failing to "promote the common good." The growth of gross domestic product (GDP) does not result in improvement in livelihood or employment rates. "The individual pursuit of rational self-interest does not promote rational social outcomes...An economic system whose existence is predicated on the exclusion of most human beings from effective membership in society cannot promote the common good. In order for an economic system to promote the common good, it must eliminate conflicting class interests; and to do so it must eliminate class distinctions" (Ionescu, 2012).

Extensive research findings demonstrate "the transmission of disadvantage across generations." Studies further reveal an association between childhood well-being and poverty. Childhood poverty is further correlated with lower test scores and educational outcomes. Accelerated testing at educational settings has been shown to have "detrimental impact" on the self-esteem of a child if he or she is already underachieving. A number of studies establish a link between academic performance and self-esteem. Studies have also demonstrated an association between delinquency and academic performance and thus employment opportunities. A strong link has been established between "class of origin and class of destination," education being the primary mediating factor. Unemployment is further related to dysfunctions in social life, including marital problems as well as psychological distress. It has been demonstrated that separation and divorce could possibly have a negative impact on children's self-esteem. Parental guidance is positively associated with childhood educational outcomes while childhood delinquency is associated with decreases in educational orientation. Children raised in households that struggle with finances are also less likely to have increased academic orientation and achieve educational success (Tomlinson & Walker, 2010).

24.6 U.S. Immigrants

Low-income legal U.S. immigrants may be particularly prone to hunger. The Personal Responsibility and Work Opportunity Reconciliation Act of 1996, also known as welfare reform, denied Medicaid, food stamps, supplemental security income and temporary assistance to legal U.S. immigrants on the grounds of their immigration status and date of entry into the U.S. Such restrictions were projected to amount to $23 billion or approximately half of the federal savings expected to be generated by the welfare reform despite the fact that legal U.S. immigrants represented only 12% of the population with incomes below the poverty threshold (Kasper et al., 2000).

Discussion

"Downward Acceleration" firmly establishes the fact that wealth is unequally distributed amongst humanity which is not surprising taking into consideration the corruption taking place amongst governmental officials, industrial and governmental ability to create problems (i.e., adding fluoride to water), to be secretive (i.e., nano technologies; fracking), deceitful (i.e., electrical vehicles), violent (i.e., oil companies), to establish regulations, policies and/or procedures preventing growth of small businesses (i.e., zoning regulations; agricultural dependency; farmers mis-treatment, etc.) while promoting expansion of corporations (i.e., nano technology) and to privatize natural resources (i.e., water) - all of which translate into the acquisition of centralized governmental and industrial profit and power.

References

Asgary, R., Garland, V., Jakubowski, A., Sckell, B. (2014). Colorectal cancer screening among the homeless population of New York City shelter-based clinics. *American Journal of Public Health*, *104*(7), 1307-1313.

Alom, F., Ward, B. D. & Hu, B. (2013). Macroeconomic effects of world oil and food price shocks in Asia and Pacific economies: Application of SVAR models. *OPEC Energy Review*, *37*(3), 327-372.

Chang, S. (2013). Impact of 2008 global economic crisis on suicide: Time trend study in 54 countries. *Nursing Standard, 28*(8), 19.

Davies, J. B., Sandstrom, S., Shorrocks, A. & Wolff, E. N. (2010). The level and distribution of global household wealth. *The Economic Journal, 121,* 223-254.

Franks, T. & Cleaver, F. (2007). Water governance and poverty: A framework for analysis. *Progress in Development Studies, 7*(4), 291-306.

Ionescu, L. (2012). Corruption, unemployment, and the global financial crisis. *Economics, Management, and Financial Markets, 7*(3), 127-132.

Kasper, J., Gupta, S. K., Tran, P., Cook J. T. & Meyers, A. E. (2000). Hunger in legal immigrants in California, Texas, and Illinois. *American Journal of Public Health, 90*(10), 1629-1633.

Tomlinson, M. & Walker, R. (2010). Poverty, adolescent well-being and outcomes later in life. *Journal of International Development, 22,* 1162-1182.

Chapter 25

MORALITY

Overview

"Morality" discusses the governance of environmental law during times of war. Specific examples of the effects of nuclear weapons and conflict (i.e., September 11, 2001) on the environment and environmental conservation efforts are presented.

25.1 Governance

The Stockholm Declaration of 1972 established the foundational principles of international environmental law which, to the most part, did not include protecting the environment during times of war. "When States resort to armed conflict they cause damage to habitats and ecosystems, contaminate soil and water, and destroy wildlife." It is thereafter rather contradictory that the international sector has made a decision to protect the environment via treaties such as the World Heritage Convention (WHC), the Convention on Biological Diversity (CBD), the Ramsar Convention on Wetlands (Ramsar) and the United Nations Convention on the Law of the Sea (UNCLOS). The UNCLOS is the only treaty that makes a reference to military activities or armed conflicts. The Rome Statue of the International Criminal Court (ICC) of 2002 has the authority to try individuals for "the most serious crimes of international concern" such as genocide, war crimes and crimes against humanity. The Rome Statute of the ICC further criminalizes the "willful infliction of 'widespread, long-term and severe damage to the natural environment.'" The scope of the ICC extends only to States which are parties to the statute. The United States (U.S.) is not a party to the statute. At present, the U.S. "is not afraid to use its formidable force where it sees it" (Bunker, 2004).

25.2 Nuclear Weapons

Efforts to outlaw weapons of mass destruction or to place them under international ownership were undertaken for over a century, yet to no avail. By the 1950s, the Soviet Union and the United Kingdom (UK) had tested their nuclear weapons, followed by France and China in the 1960s. Between 120,000 and 240,000 people or between 20% and 50% of humanity exposed to the nuclear bombs dropped at Hiroshima and Nagasaki in 1945 died within 4 months of the disaster. Leukemia, growth retardation, arteriosclerosis and solid cancers were longer-term causes of death resulting from the dropping of the nuclear bombs. In reference to "nuclear winter," studies "have postulated the possibility of dire global climate change impacts that could occur in event of a regional nuclear war." Nuclear test treaties that prohibit atmospheric nuclear tests have drastically reduced environmental radioactive contamination. A review of studies investigated the radioactive dose emitted from the U.S. nuclear test sites between 1950s and 1960s. Findings reveal that 49,000 thyroid cancers resulted from above-ground testing fallout-related radiation released from the Nevada test site between 1951-1957. 1,800 deaths resulted from fallout-related external and internal radiation exposure. Nuclear energy is increasingly perceived as a low-carbon energy source. The Chernobyl and Fukushima nuclear plant disasters represent the "grave local, regional, and global impacts of a major accident" (Dreicer & Pregenzer, 2014).

25.3 September 11, 2001

Following the terrorist attacks of September 11, 2001, section 102 of the REAL I.D. Act of 2005 was passed to allow the

Secretary of Homeland Security to "waive all laws he deems necessary" to build a border fence between the U.S. and Mexico as a national security measure. Environmental groups are concerned that Section 102 compromises "environmental integrity" and environmental laws (Jones, 2009).

International collaboration is imperative in addressing global conservation challenges. "The increasingly fortified barrier" between the U.S. and Mexico as a result of the terrorist attacks of September 11, 2001 "fragments species ranges and severs wildlife migrations...and is a metaphor for what separates humanity." Such altered borderline reality has contributed to diminished opportunities of U.S. and Mexico conservation biologists to collaborate. Political tensions between the U.S. and Mexico, drug conflicts, intensified border control, increasing cultural distrust and a number of additional barriers have contributed to a substantial reduction of collaborative projects between U.S. and Mexico conservation biologists. Meeting conservation objectives of the region requires bi-cultural collaboration between the U.S. and Mexico. "Meaningful international collaboration starts with dismantling the barriers that impede scientists from meeting one another" (Wilder et al., 2013).

Discussion

War provides the greatest evidence for the fact that profit and power are the top priority of governments and industries. If governments and industries are willing to go as far as to inflict human and environmental violations for purposes of obtaining oil, commercially release toxic substances for profit, etc., there is no reason not to confidently deduce that conflicts preceding wars are deliberately orchestrated (i.e., problem) as a means to acquire profit and power. The WWII as part of "GMO Invasion" depicted an ideal example of how conflict and war resulted in immediate profit and power for GM companies.

References

Bunker, A. L. (2004). Protection of the environment during armed conflict: One gulf, two wars. *Reciel, 13*(2), 201-213.

Dreicer, M. & Pregenzer, A. (2014). Nuclear arms control, nonproliferation, and counterterrorism: Impacts on public health. *American Journal of Public Health, 104*(4), 591-595.

Jones, R. R. (2009). Risky business: Barriers to rationality in Congress. *Ecology Law Quarterly, 36*, 467-497.

Wilder, B. T., O'Meara, C., Narchi, N., Narvaez, A. M. & Aburto-Oropeza, O. (2013). The need for a next generation of Sonoran Desert researchers. *Conservation Biology, 27*(2), 243-245.

Chapter 26

PLANET DE-CONSTRUCTION

Overview

"Planet De-Construction" presents examples of endangered, threatened and extinct animal and plant species and ecosystems. Laws and regulations pertaining to endangered and threatened species are also elaborated upon. "Planet De-Construction" discusses conservation areas and their governance, habitat change, species migration and biodiversity.

***

Philosopher Charles Taylor proposes that "changing patterns of thinking" in Western culture altered our view of the environment and of the natural (Schosler et al., 2013).

26.1 Endangered & Threatened Species

More than 1500 mammals, reptiles, amphibians, birds and plant species as well as more than 3000 ecosystem types are currently being threatened with extinction (Ritchie et al., 2013).

26.1.1 Competing Stakeholders Objectives

A great number of stakeholders with an array of conflicting objectives influence decision-making processes regarding conservation of endangered and threatened species. Economic, environmental, political and public factors shape stakeholders' objectives. Objectives "may reflect politics and profitability as much as science." Ambiguity between factual and value-based arguments, poor identification of key conservation objectives and inadequate allocation of resources result from competing stakeholders' objectives (Gregory et al., 2013).

26.1.2 Conservation

Translocation of endangered species in the United States (U.S.) from wild populations or breeding setting (i.e., zoos) has played an essential role in the recovery of endangered populations. Translocation has been utilized to reintroduce species to regions where they have been eradicated and to increase genetic diversity within isolated species populations demonstrating signs of inbreeding (Finseth & Conrad, 2014).

Reliable estimates on population density and size are elemental in the successful management and recovery of endangered species. Yet, such information, typically concerning large carnivores, is challenging to obtain and therefore not available for a large number of the world's most endangered species (Sollomann et al., 2013).

Threatened and endangered species are characterized by a reduction in genetic diversity. Therefore, the main objective of captive breeding programs of endangered species is to preserve genetic diversity and prevent species extinction (Hammerly et al., 2013).

Factor, contributing to species decline and extinction, is apparent competition which is defined as the "indirect interaction between 2 or more prey species through a shared predator." Some scientists propose active management of alternate prey, predators or both to preserve species impacted by apparent competition. Active management is controversial for three main reasons: (1) public opposition to active management; (2) active management may not have long-term effects if the causes of the decline such as habitat change are not addressed; (3) the removal of alternate prey and predator species can have an unpredicted negative impact on prey species subjected to apparent competition. Aside from being controversial, active management geared towards the preservation of species affected by apparent competition may be "affected by social values and politics rather than science alone." Ethical ambivalence could also arise from deciding which species to remove in ecosystems where both prey and predator species are native. Policy makers are "ultimately" responsible for implementing management strategies. They take public expectations, policies, social and political values under consideration prior to making a management decision. Therefore, instances when management decisions are in opposition with science-based conservation suggestions are projected to take place (Wittmer et al., 2012).

26.1.3 Endangered Plant & Tree Species

26.1.3.1 Plant Species

The International Union for Conservation of Nature Redlist (IUCN Redlist) has estimated that there are approximately 9,100 endangered and threatened plant species in the world. Endangered plant species are exposed to a number of threats including small population sizes, habitat fragmentation, invasion of alien plant species and changes in pollinator abundances. The state of Utah has more than 300 rare plant species, 24 of which are federally listed as either threatened or endangered (Lewis & Schupp, 2014).

26.1.3.2 Tree Species

Montane bamboo which grows in tropical Africa and Asia is of great importance to people inhabiting the regions adjacent to Echuya Forest Reserve in West Uganda. Bamboo trees are used for building poles, weaving baskets and bean-staking. Bamboo trees in Echuya have been declining due to poor harvesting techniques, damages resulting from insect borers and the invasion of alien forest trees. Climate change may also play a role in the decline of bamboo trees in Echuya. Temperatures have been increasing in the past 30 years in most areas of south western Uganda due to "clearings of forests and swamp drainages." 70% of the local people interviewed agreed that over-harvesting is the main reason for the decline of bamboo trees in Echuya. Provided bamboo forests in Echuya continue to decline at the present rate, biodiversity and the livelihood of people inhabiting the regions surrounding the Reserve could be compromised (Bitariho & McNeilage, 2007).

26.1.4 Endangered Animal Species

26.1.4.1 Giraffe

The population of the Thornicroft's giraffe, Giraffa camelopardalis thornicrofti, is isolated and limited to Zambia's South Luangwa Valley (Fennesy et al., 2013).

26.1.4.2 Wolf

The wolf (Canis lupus) was hunted to extinction in Sweden in the early 20th century. It is currently considered an endangered species by Swedish authorities (Laikre et al., 2012).

26.1.4.3 Polar Bear

Polar bears are "a circumpolar Arctic species" (Pagano et al., 2012). Polar bears (Ursus maritimus) are being deprived of their natural habitat due to climate-induced changes in sea ice structure, spread, break-up and freeze-up patterns. Polar bears rely on sea ice as a hunting platform. "Multiyear ice cover in the Arctic Ocean has decreased from about 75% in the mid-1980s to 45% in 2011" (Stirling & Derocher, 2012). Reduction of sea-ice in recent years has been associated with reduction in polar bear population size and survival (Pagano et al., 2012). Polar bears are projected to disappear from the southern regions of their range within the next 30 to 40 years provided the climate continues to warm, eliminating sea ice. A different estimate suggests that about two thirds of the world's polar bears would be extinct by the middle of the century. Human settlement in the Arctic, food web contamination, industrial activities, reduced prey species could add to the effects of climate warming, "posing a profound threat to polar bear survival." In 2008, polar bears were listed as "vulnerable" by the IUCN Red List and "threatened" under the U.S. ESA (Stirling & Derocher, 2012).

26.1.4.4 Blue Antelope

Habitat degradation and loss, SLR, competition, geographical isolation and excessive hunting by European colonists are considered contributing factors to the extinction of the blue antelope, Hippotragus leucophaeus, in southern Africa in c. AD 1800 (Faith & Thompson, 2013).

26.1.4.5 Rare Wild Animal Species

Rare wild animal populations are increasingly being threatened by infectious diseases. A large number of rare wild animals were vaccinated in the past to prevent disease outbreaks. Due to financial and time restrictions characterizing large-scale disease management via large-scale vaccination, restricted number of rare wild animals are vaccinated in a prophylactic manner to prevent extinction of the species rather than disease outbreak. "Many wild populations will require continuing management, potentially in perpetuity, to avoid extinction" (Doak et al., 2013).

26.1.4.6 Emperor Penguin

Emperor penguin populations along the west coast of the Antarctic Peninsula declined from 250 breeding pairs in the 1970s to 20 breeding pairs in 1999, culminating in extinction of the specie in 2009. Current simulations project a decline of the emperor penguin towards extinction at Terre Adelie, in Antarctica provided the climate continues to follow simulated by the present study patterns by the end of the century (Ouvrier et al., 2012).

26.1.5 Endangered Bird Species

Over 40 bird species have become extinct in New Zealand since the arrival of human populations (Horrocks et al., 2008).

"Grasslands are among the most threatened habitats in North America." The number of many grassland bird populations is decreasing. For example, the populations of the Short-eared Owl and the Northern Harrier have declined in the past decades. Essential factors such as fire regimes, transformation of grasslands to croplands and grassland fragmentation contribute to the decline of grassland bird populations (Vukovich & Ritchison, 2008).

Prescribed burning is one management activity undertaken in the U.S. to maintain and improve the habitat of the endangered since 1973 red-cockaded woodpecker (RCW) species (Finseth & Conrad, 2014).

26.2 Regulation

26.2.1 CITES

The Convention on International Trade in Endangered Species (CITES) regulates the international trade of endangered species. Internet wildlife trade is a new form of endangered species trade. The present study examined the status of e-commerce of CITES-listed species. 24 cacti sellers were monitored over a period of 6 months. 1000 cacti listed in

Appendix I have been sold during that period of time. CITES permits were issued to approximately 25% of the cactus plants being sold. "The potentially wide scale of the illegal global trade...should raise concerns about the adequacy of the protection for CITES species" (Sajeva et al., 2013).

26.2.2 U.S. Endangered Species Act (ESA)

The U.S. Endangered Species Act (ESA) was enacted in 1973 (Regan et al., 2013) as a means to "prevent species' extinction and promote their recovery." The ESA is considered as one of the most significant conservation tools in the U.S. (Paeth et al., 2009) and "one of the most powerful and influential environmental laws in the United States" (Regan et al., 2013). Plant, vertebrate and invertebrate species are protected under ESA. One of its greatest strengths is its "reliance on robust science to inform decision making."

The U.S. Fish and Wildlife Service (USFWS) is an agency that is part of the Department of the Interior and the National Marine Fisheries Service (NMFS) is part of the Department of Commerce. Both USFWS and NMFS share responsibility for implementing the ESA. USFWS is responsible for freshwater and terrestrial species while NMFS is entitled to protecting anadromous and marine species (Paeth et al., 2009).

26.2.2.1 Listing

The U.S. ESA allows for the listing of species as either endangered or threatened. Species are listed either as endangered or threatened under ESA only if they meet provisional standards, definitions and categories. An endangered species is an organism "in danger of extinction throughout all or a significant portion of its range" while a threatened species is an organism that "is likely to become an endangered species within the foreseeable future throughout all or a significant portion of its range." "Unfortunately most conservation measures contain ambiguous terms, such as the ESA's foreseeable future and significant portion of its range." Ambiguity prevents an appropriate application of risk and scientific analysis relative to species proposed for ESA listing (Paeth et al., 2009). There are no policy guidelines on interpreting the "vague definitions of endangered or threatened." Neither are there guidelines on the degree of extinction risk as a prerequisite to species protection. Such decisions are made on a case-by-case basis, "leaving the process vulnerable to being considered arbitrary or capricious" (Regan et al., 2013). Species that have not been listed under ESA are not protected by ESA. Only limited protection to species is afforded once species are proposed for ESA listing. 361 mammals, 126 reptiles, 35 amphibians, 166 fishes, 317 birds, 241 invertebrates and 856 plants have been listed under ESA. The ESA funds and authorizes actions that do not endanger enlisted species' continual existence or natural habitat (Paeth et al., 2009).

26.2.2.2 Recovery

Recovery plans are the primary instrument for guiding species recovery under the ESA. Recovery plans "identify priority conservation actions, steer recovery work, provide measurable recovery criteria, and inform management agencies, elected officials, and the public as to the status and conservation needs of imperiled species" (Povilitis & Suckling, 2010). The listing agency is responsible for the development and implementation of a recovery plan for the species being listed under ESA. ESA requires USFWS and NMFS to develop and implement recovery plans. USFWS and NMFS do not have the authority to ensure that other federal or non-federal agencies comply with the recovery plan (Paeth et al., 2009).

Population viability analyses (PVAs) are tools designed to establish recovery criteria and strategies for threatened and endangered species under the ESA. "PVAs, especially as

379

currently implemented, are not sufficiently robust for setting quantitative endangered species recovery criteria" (Zeigler et al., 2013).

In 2007, the U.S. Fish and Wildlife Service (FWS) removed the Bald Eagle (Haliaeetus leucocephalus) from the ESA threatened species list upon the restoration of the Bald Eagle across its historic range. On the other hand, the gray wolves (Canis lupus) were declared recovered in five states in Western U.S. in 2009 despite the fact that the species was absent "from the majority of its former range in these states." These examples "portray...the arbitrary nature of standards that form the basis of both listing and recovery determinations under the statute." One of the purposes of ESA is to conserve "the ecosystems upon which endangered species and threatened species depend." The language of ESA therefore implies that ecosystems are to be protected by ESA, "not for their own sake," but only when it is necessary to do so for the conservation of species (Carroll et al., 2010).

Addressing climate change in environmental conservation planning is essential. Yet, "the management and legal context of particular efforts can place constraints on how supporting science is conducted and used" (McClure et al., 2013). Climate change has been increasingly integrated in recovery plans in the past decade, yet it has not been fully incorporated. More than 87% of the 1209 recovery plans developed until 2008 did not integrate climate change factors. 59% of recovery plans between 2005 and 2008 integrated climate change factors (Paeth et al., 2009). It is imperative that recovery plans for species "acutely threatened by climate change" include recommendations for reducing GHG emissions "as a matter of urgency" (Povilitis & Suckling, 2010).

The ESA has helped to prevent further harm to and extinction of endangered species, and yet "it has largely failed" to contribute to the recovery of endangered species populations. One reason is the failure of EPA to implement recovery programs that will allow endangered species to be removed from the Endangered Species List. Despite EPA's mandate that the Department of Inferior devise a recovery plan for each endangered species, this requirement has only been partially implemented (Gersen, 2009). "Species listed under the U.S. Endangered Species Act (i.e., listed species) have declined to the point that the probability of their extinction is high" (Leidner & Neel, 2011).

26.2.2.3 Cost

The cost for species recovery in the U.S. is substantial. The 5-year (2008-2012) plan for the recovery of the Florida panther, for example, was projected to cost $17.75 million. The recovery plan for red wolf has cost $1 million per annum since 1974 while the annual cost to save the California condor amounts to $5 million (Finseth & Conrad, 2014).

26.2.2.4 Take Permit

Harming, harassing, killing, injuring, taking into one's possession or pursuing the majority of species listed under the ESA "is prohibited unless specifically authorized via a case-by-case permit process" (Paeth et al., 2009). ESA exemption was established in response to "human economic interests." It allows for "possible extinction of a species" in order to accommodate industrial interest. Agencies, including governmental agencies, are exempted from ESA requirements if the benefits of proposed by agencies projects supersede the benefits from conserving species (Yuknis, 2011). Take permits are granted by USFWS and NMFS to nonfederal agencies "for scientific purposes or to enhance the propagation or survival." Incidental take permits (ITPs) are authorized for activities including "commercial and recreational fisheries, water diversion, forest management, and land development." ESA agencies are required to utilize the "best available scientific...and commercial information" during key decision-making processes (Paeth et al., 2009).

26.2.3 California Endangered Species Act (CESA)

The standard required to enlist a species as endangered under the California Endangered Species Act (CESA) is defined as the "amount of information...that would lead a reasonable person to conclude that there is a substantial possibility the requested listing could occur" (Cheng, 2009).

26.2.4 Executive Order (EO) 13045: Protection of Children from Environmental Health Risks and Safety Risks

In 1997, President Clinton signed Executive Order (EO) 13045, Protection of Children from Environmental Health Risks and Safety Risks. EO 13045 directs federal agencies to identify and assess environmental health and safety risks that disproportionately affect children and to address such risks within their policies, standards, programs and undertakings. The U.S. EPA established the Office of Children's Health Protection to assist in the implementation of EO 13045. The present study evaluated proposed and final rules or regulations published by the EPA from 1998 until 2006. Findings revealed that 99% of the 1,658 reviewed regulations discussed EO 13045. Yet only a small percentage of regulations are "economically significant" and are hence "subject to its requirement to evaluate children's health risks." EO 13045 does not spread across a broad enough range of regulations to afford children the protection that they are entitled to (Payne-Sturges & Kemp, 2008).

26.3 Species Extinction

"Over the last 100 years, 785 species, including mammals, seabirds, fishes, crustaceans, and gastropods, have become extinct or are at the edge of extinction." Habitat loss, overexploitation, disease, pollution and climate change are proposed as causative to such "alarming phenomenon" factors (Wang & Zhou, 2013).

The demise of the last individual of a species is indicative of the unrecoverable "loss of a unique form of biodiversity." Since extinction is seldom witnessed, it must be inferred (Collen et al., 2010). The number of endangered and extinct species is growing. Five primary contributing to extinction factors have been proposed – alien species invasion, species over-exploitation, climate change, habitat loss and co-extinction (Dunn et al., 2009). Geographic isolation is a powerful predictor of extinction (Bennett & Arcese, 2013). At least 23 birds, 27 mammals, 4 frogs and more than 60 plant species have become extinct in the last 200 years in Australia (Ritchie et al., 2013). East Greenland arctic wolf populations were poisoned and exterminated by Norwegian and Danish hunters between 1899 and 1939 (Marquard-Peterson, U., 2012). 80% of all species extinctions in the world have taken place on islands. Approximately 2000 bird species on tropical Pacific islands have become extinct since human arrival on the islands (Wetzel et al., 2013).

26.3.1 Islands

Small oceanic islands house a variety of threatened species. The majority of extinct species since the 1,500 AC inhabited small oceanic islands. Natural disasters in addition to damaging human activities could have detrimental impact on the biodiversity of islands.

Natural disasters contribute to biodiversity and ecosystem damage. Tsunami events in relatively small oceanic islands could contribute to the extinction of species as well as to the degradation of soil cover which prevents the growth of plants. When natural catastrophes are combined with damaging human activities, "their effects could become devastating."

Detrimental human activities are the main cause of ecosystem degradation in oceanic islands (Hahn et al., 2014). "Human activity is generally considered the prime cause of risk of extinction for the vast majority of species" (Loehle & Eschenbach, 2012). Human activities on oceanic islands such as hunting, habitat degradation and the introduction of predators, herbivores and competing species have contributed to the extinction of species (Hahn et al., 2014). Deforestation also plays an important role in species extinction. Climate change is "assumed to be a future risk factor" (Loehle & Eschenbach, 2012).

Loehle & Eschenbach (2012) investigated the historical extinction rates of mammals and birds in islands (including Australia) and continents. 95% of the extinction of birds and mammals since 1500 occurred on islands. 6 continental birds and 3 continental mammals were recorded as extinct since the 1500 while 123 island birds and 58 island mammals were recorded as extinct since 1500.

26.3.2 Species Co-Extinction

Co-extinction or "the loss of one species as a result of the extinction of a species it depends on," results in a cascading effect along trophic levels. Models propose that co-extinction is the most widely spread form of biodiversity loss. A mass co-extinction event or many thousands of co-extinctions are predicted by models to be taking place at the moment. Yet, "paradoxically," only a few past or present co-extinction events have been actually documented. Mutualistic and parasitic interactions are expected to be most severely affected by such effects. Co-extinction "rarely acts alone." Rather, it is influenced by the remaining major extinction factors, namely climate change, over-exploitation, species invasion and habitat loss. Global climate change has contributed to "early warnings of biological responses to warming," including changes in plant phenology, bird migration and reproduction and in emergence times in insects.

"The extinction crisis is dire and conservation resources are scarce" which may contribute to the argument that conservation efforts concerning parasites should be ignored. The extinction of parasites results in consequences as great as those pertaining to the extinction of their hosts. It is likely that millions of parasite and mutualists are already extinct, though the actual number of extinction events is unknown. The extinction rate of vertebrates, including mammals or birds, is also uncertain (Dunn et al., 2009).

26.3.3 Finland

The International Union for Conservation of Nature (IUCN) Red List Index (RLI) is one of the most comprehensive national assessments of species in Finland. The RLI reveals trends in the conservation status of particular species as an indication of the trends in biodiversity in Finland. Taxonomic groups in Finland include a variety of organisms from birds and mammals to lichen, moss and vascular plants. The present study investigated Finland's RLIs for 11 taxonomic groups of organisms. Findings revealed that the rate of extinction of the analyzed taxonomic groups was higher in 2010 compared to 2000 (Juslen et al., 2013).

26.3.4 Environmental Filtering

According to the concept of environmental filtering, "the environment acts as a filter that favours species characterized by certain traits over species with different traits" (Hanspach et al., 2012).

26.4 Biodiversity

Biodiversity is greatest in developing countries (Naughton-Treves et al., 2005). "Genetic diversity is important for the maintenance of population viability and adaptive potential of species" (Ethier et al., 2012). "The international goal of substantially reducing the loss of biological diversity and the effects of this loss by 2010 was not achieved" (Juslen et al., 2013).

26.4.1 Biodiversity Conservation

Biodiversity conservation plays an important role in commodity production landscapes. For example, pollination insects or plants that decontaminate waters in wetlands are tremendously valuable in production landscapes. "Many land users accept some responsibility for severely transforming the earth's surface."

Biodiversity awareness is growing in South Africa; more students are enrolling in biodiversity conservation and environmental management courses and degrees (Crous & Roets, 2014).

26.4.2 Nitrogen (N)

Nitrogen (N) is a fundamental element of Earth's atmosphere, biosphere and hydrosphere (Chen et al., 2010). Unlike the majority of pollutants, N is an essential nutrient which, in non-toxic levels, contributes to the growth of plant species (Johansson et al., 2012). Wild and prescribed fires contribute to the loss of N within tropical ecosystems (Chen et al., 2010). Short-term exposure to elevated N concentrations exerts toxic effects on lichens (Johansson et al., 2012) and contributes to chlorophyll deterioration and membrane damage (Chen et al., 2003). N deposition has increased on a global scale in the past 150 years. N deposition increase is considered to be one of the primary threats to biodiversity in the world (Johansson et al., 2012).

26.5 Environment

Human beings have been altering the environment since "we gained control over fire." In the Holocene epoch, "even prior to the expansion of Europe and industrialization," "no parts of the planet remained pristine" except for Antarctica (Kirch, 2005).

26.5.1 Osborn & Vogt

Fairfield Osborn published the bestselling book on environmental challenges titled *Our Plundered Planet* in 1948. William Vogt published a different bestselling book concerning environmental problems in 1948 called *Road to Survival*. Both authors "warned about a concern – the depletion of natural resources." According to the authors, "the central work of conservation was to reduce violence." Osborn contended that man's separation from the land and his or her "apathy" towards the use of natural resources stems from the fact that 55% of the human population lives in cities and towns. Both authors deduced that the scarcity of natural resources, overconsumption, and environmental degradation have contributed to the global wars of their time. According to Osborn, "ecological imbalances...threatened 'man's very survival.'" Both Osborn and Vogt believed that "the new world order based on the spread of industrialization and growth-based economies would yield not peace but greater instability and conflict." Both authors were further concerned about the environmental degradation compromising Earth's carrying capacity, especially in regards to soil erosion. In the opinion of Osborn, society was obliterating "the resources it depended on", placing its own existence at risk. "The uncomfortable truth," according to Osborn, "is that man during innumerable past ages has been a predator – a hunter, a meat eater and a killer" (Robertson, 2012).

26.5.2 Environmental Science

Policy-makers and specialists were the only professionals who addressed environmental challenges prior to the 1960s. They were perceived as "eccentric alarmists." There were "two environmental revolutions" since the 1960s. The first environmental revolution took place between the late 1960s and the early 1970s. It concerned economic growth in relationship to environmental quality. The second revolution took place from the 1980s and it focused on discussions regarding accomplishing economic growth in environmentally beneficial manner. The current assumption is that properly managed economic development would contribute to environmental growth and that solid environmental policies would contribute to economic development (O'Keeffe, 2009).

"Over the past 15 years, following on from the Convention on Biological Diversity (CBD), environmental sciences and research in general have gained significant conceptual and financial support" (Klopper et al., 2006).

26.5.3 Urban Environmental Concerns

International cooperation to address environmental concerns is paramount in light of exponentially growing urbanization. Developing countries are increasingly being challenged by growing urbanization trends and lack of adequate environmental management (Memon et al., 2005).

26.5.3.1 Asia

Approximately 30% of the population in Asia resided in cities in 2000. A wide variety of environmental issues are predominant in urban areas in Asia, including air quality and solid waste management, transportation, sanitation and water supply. The three fundamental sources of air pollution are also the "pillars of economic development." They are construction, industry and automobiles. City planning is inadequate in Asia, leading to a "lack of sustainable solutions" to urban environmental problems. More than 1 kg of solid waste is generated daily per person in developed cities and about half of that amount daily per person in developing cities. Urban environmental concerns are projected to increase unless measures are undertaken to counteract such increase (Imura et al., 2005).

26.5.3.2 China

Rapid population and economic growth, mass migration, coastal development and loss of habitat are prevalent in China. China's accelerated economic growth has "exacerbated" a number of wicked environmental problems. Wicked problems in environmental and social planning context are characterized by "inadequate governance, missing institutions, and shortage of time before the problem becomes even harder to address." Wicked problems involve a variety of "players" such as governmental and non-governmental organizations and industries. Conflict among players typically arises from the attempt to find an optimal solution to a problem. "Policies that advantage one set of players often disadvantage others, creating winners and losers." Partial solutions to wicked problems often contribute to the development of new problems (Hughes et al., 2012).

26.5.4 Ecosystems

"With the alarming loss of natural habitats around much of the world, humanity is being robbed of essential ecosystem services such as air purification, weather regulation, maintenance of soil fertility and stability, waste detoxification and pest control" (Bradshaw et al., 2007).

The ultimate objective of the practice of land-use change is "the acquisition of natural resources for immediate human needs usually at the expense of environmental degradation" (Fisaha et al., 2013). Habitat fragmentation, chemical pollution and changes in community structure are the major human activities that cause alternations of ecosystems. The focus of chemical pollution in the last 3 decades has been on three "priority" contaminants, including lead (Pb) and mercury (Hg) (Kreisberg, 2007). The ambiguity surrounding ecosystem regulating services contributes to ecosystem overexploitation and degradation (Crafford & Hassan et al., 2014). Management of ecosystems requires an observation and understanding of ecosystem dynamics as a means to respond to changes within the environment (Folke et al., 2005).

26.5.5 Forests

Forests represent 90% of the terrestrial biomass (Bader et al., 2013). Long-term forest conservation is threatened by "agricultural expansion, infrastructure development, and resource extraction" resulting from growing human population. For example, in spite of biological diversity losses, substantial forest areas are cleared each year in India for purposes of agricultural expansion (Krishnadas et al., 2013). India is one of the top 12 mega-biodiversity countries, containing 6% of the forests of the world. India's forest cover is 692,027 km^2 which accounts to 21% of the total geographical area in India (Krishna et al., 2014). As of 2003, about 15% of the rain forests of Brazil have been converted into managed landscapes (Bagley et al., 2014). Approximately 15.2 million hectares of tropical rainforests were lost in the 1990s. 90% of lost tropics were converted to alternative land uses (Yonekura et al., 2012). Converting forests to agricultural lands is "the most extreme form of forest disturbance" (Nepstad et al., 2006).

200,000 hectares of coastal land in southeastern U.S. are occupied by tidal forests. "Tidal forests and their delivery of ecosystem services face a tenuous future unless they can migrate upriver, and that is unlikely in most areas because of topographic constraints and increasing urbanization of the coastal zone" (Craft, 2012).

"Prescribed burning is nowadays the major cause of forest fires which can spread over wide areas during drought" (Rius et al., 2009). Wildfires "consume millions of acres of forest each year" (Mauderly & Chow, 2008).

26.5.5.1 Deforestation

Fragmentation and destruction of tropical forests are the primary reasons for the extinction of species in the world. Fragmentation of tropical forests contributes to reduced species population size as well as to increased risk of species extinction as a result of immediate causes such as fires, hurricanes and disease outbreaks (Ribon et al., 2003).

Deforestation is defined as "an event in a geographic area associated with a change in land use from forest to non-forest category" (Krishna et al., 2014). Deforestation may restrict the movement of a number of forest species and contribute to an increase of the distance to climate analogs by substituting areas that would have otherwise been appropriate climate analogs with "human-dominated habitats." The unwillingness of seed dispersers to cross deforested areas may further prevent seed dispersal in such regions and therefore movement to climate analogs. A number of forest inhabitants including birds, insects and mammals are unwilling or unable to cross deforested areas, posing challenge to organisms' successful migration to climate analogs. Under Business-As-Usual (BAU) simulated deforestation projections, 75% or more of seven ecoregions in tropical South America have no climate analogs. A large number of plant and animal species in these ecoregions are projected to be lost "as they are unable to respond to climate change through shifts in their distributions to areas with comparable climates" (Feeley, et al., 2012). Deforestation contributes to loss of biodiversity

391

(Kartzinel et al., 2013).

The historical fire events on South Island, New Zealand, that contributed to rapid deforestation and erosion rates of the region were investigated. Results reveal that not a single, but multiple wildfires (reburns) driven by Polynesian populations may have taken place 700-800 years ago on South Island, New Zealand (McWethy et al., 2009).

7% of all Atlantic forests has remained, less than 1% of which is at an intact state. The rapid levels of destruction and fragmentation of tropical forests may contribute to a great number of bird species becoming rapidly extinct (Ribon et al., 2003).

Habitat fragmentation (i.e., deforestation) in south-central Africa is the direct result of an increasing human population (Staub et al., 2013).

26.5.5.1.1 Congo

Population growth in the Congo basin is projected to contribute to a 100% increase in demand for agricultural land in the next 2 decades. The resulting deforestation is expected to lead to 0.78 degrees C^0 increase in surface temperature at the Congo basin by the middle of the 21st century (Akkermans et al., 2014).

26.5.5.1.2 Haiti

The forest area in Haiti represents 3% of its total area. Haiti suffers from high deforestation rates. Its forest areas declined by 50% between 1990 and 2000. Moreover, between 10,000 and 15,000 hectares of fertile land is lost to soil erosion every year in Haiti (Conceicao & Mendoza, 2009).

26.5.5.1.3 India

The present study investigated deforestation patterns in

Andhra Pradesh, India, between 1930 and 2011. Andhra Padesh has the second largest forest area amongst all states in India. There were 3,981 forest patches in 1930, 5,553 in 1960, 8,760 in 1975, 9,412 in 1985, 9,646 in 1995 and 10,597 in 2011. The mean patch size of forests declined from 21.5 km² in 1930 to 12.3 km² in 1960, culminating at 3.9 km² in 2011. The forest cover accounted to 85,392 km² in 1930, 68,063 km² in 1960, 46,940 km² in 1975, 45, 520 km² in 1985, 44, 409 km² in 1995, 43, 577 km² in 2005 and 43, 523 km² in 2011. The annual rate of deforestation between 2005 and 2011 was 0.02% (Krishna et al., 2014).

26.5.5.1.4 Amazon

Large-scale deforestation in the Amazon has only been taking place in the last few decades (Medvigy et al., 2013). As of 2002, about 15% of the Amazon rainforest "had already been lost to deforestation." Deforestation in the Amazon is expected to increase to almost 50% by 2050. According to conservative estimates, 25% of Amazonian forests are projected to be lost (Feeley, et al., 2012).

Deforestation contributes to the spread of malaria, a major public health threat in the Brazilian Amazon. Forest fires, selective logging as well as road construction increase malaria risk. Annual deforestation rate in the Amazon ranges between 6,000 and 28,000 km² per year. Over 17% of pre-1970s Amazon forests have been cleared. Agriculture, gold mining and timber extraction "have transformed the forest fringes" (Hahn et al., 2014b).

26.5.5.1.5 Tropical Areas

"15%-30% of the natural forest cover has been converted to pasture or cropland." The rate of deforestation in tropical areas has increased exponentially in past decades. A large proportion of tropical forests are projected to be lost by the end of the 21st century (Davin & de Noblet-Ducoudre, 2010). Provided tropical plant species are unable to tolerate or adapt to a warming climate, "the effects of climate change will overshadow the effects of deforestation" leading to "massive losses of biodiversity" (Feeley et al, 2012).

26.5.5.1.6 Dry Rangelands

Dryland ecosystems comprise approximately 40% of the Earth's terrestrial land mass. Drylands contain biological soil crusts or biocrusts which serve the purpose of stabilizing soil surfaces, including plant performance and controlling hydrologic processes including carbon (C) and nitrogen (N) inputs. C and N stores contribute to the nutritional and overall health of soils (Johnson et al, 2012). Climate and land use change contribute to the desertification of dry rangelands such as semi-arid savannas. The result from deforestation is ground surface warming. Human population growth causes "higher stress levels" in dry rangeland systems. A number of studies agree that dry rangeland systems are currently not in a state of equilibrium (Lohmann et al, 2012).

26.5.5.1.7 Grasslands

Grasslands represent 40% of global land area (Crous et al, 2010). Tropical grasslands are essential for biodiversity, yet they are "among the most threatened biomes globally" due to their transformation into agricultural land. Vietnam and Thailand no longer encompass considerable areas of grassland due to the transformation of grassland into agricultural production areas or the construction of reservoirs associated with agriculture (Packman, 2013).

26.5.5.1.8 Climate Change & Global Warming

Deforestation and climate change are the major "human

disturbances" threatening tropical forests (Feeley et al, 2012). Deforestation contributes to 18% of global carbon dioxide (CO2) emissions (Krishna et al., 2014). Deforestation and biomass decomposition amounted to 17.3% of total GHG emissions in 2004 while fossil fuel combustion represented 56.6% (Bourgeon et al., 2012). Deforestation contributes to surface albedo increase and subsequent cooling effect on the climate. On the other hand, deforestation decreases the rates of evapotranspiration which contributes to warmer surface climate. The surface of the ocean cools as a response to deforestation (Davin & de Noblet-Ducoudre, 2010).

26.5.5.1.9 Floods

Almost 100,000 people were killed and 320 million people were displaced by floods in 56 countries between 1990 and 2000. The total economic loss exceeded $1151 billion. Each year floods kill and displace hundreds of thousands of people particularly in developing countries. Billions of dollars of damages result from flooding each year. Data collected between 1990 and 2000 further revealed that the frequency of flood was negatively correlated with the amount of natural forest left and positively associated with the loss of natural forest area. Vegetation loss can result in increased runoff due to impaired tree water evaporation and soil hydraulic conductivity. Reforestation may contribute to the reduction of the frequency and intensity of flood-related disasters. Reducing deforestation rates is projected to alleviate the frequency and intensity of flood events (Bradshaw et al., 2007).

26.5.5.1.10 Droughts

"Two once-in-a-century-level droughts" have occurred in the Amazon between 2003 and 2010. The first major drought occurred in 2005 and affected about 1.9 million km^2 while the second major drought from 2010 affected 3.0 million km^2.

Increases in the frequency of fires, reduced rates of river drainage and a reversal of the Amazon net sink of carbon resulted from the droughts. Current rates of tropical deforestation and record high tropical North Atlantic sea surface temperatures appear to have contributed to the severity of the droughts (Bagley et al., 2014).

There is "increasing occurrences of drought-and heat-inducted tree and forest mortality." Droughts have an impact on the carbon balance and hydraulics of trees (Hartmann et al., 2013).

26.5.5.1.11 Indigenous Reserves

There are growing international efforts to reduce the impact of climate change by decreasing the rates of tropical deforestation. Such efforts increasingly depend on "indigenous reserves as conservation units and indigenous peoples as strategic partners." Despite the fact that there is no supporting evidence, anthropogenic fire associated with ceremonial and hunting practices of indigenous people in the Neotropical savannas of Central Brazil is "routinely represented in public and scientific conservation discourse as a cause of deforestation and increased CO_2 emissions." The present study investigated such claim for the Xavante people of Pimentel Barbosa Indigenous Reserve in Brazil. Findings reveal that "the real challenge...is the long-term sustainability of indigenous lands and other tropical conservation islands increasingly subsumed by agribusiness expansion rather than the localized subsistence practices of indigenous and other traditional people" (Welch et al., 2013).

26.5.6 Fragmentation

Fragmentation is defined as "a process during which a large expanse of habitat is transformed into a number of smaller patches of smaller total area, isolated from each other by a matrix of habitats unlike the original" (Krishna et al., 2014). Habitat fragmentation "can affect seed production and gene flow of plants," and threaten plant biodiversity (Nagamitsu et al., 2014).

26.5.7 Habitat Loss

"With the alarming loss of natural habitats around much of the world, humanity is being robbed of essential ecosystem services such as air purification, weather regulation, maintenance of soil fertility and stability, waste detoxification and pest control" (Bradshaw et al., 2007).

The ultimate objective of the practice of land-use change is "the acquisition of natural resources for immediate human needs usually at the expense of environmental degradation" (Fisaha et al., 2013).

26.5.8 Invasive Plant Species

Habitat degradation is accompanied by non-native species invasion, resulting in environmental changes such as nutrient fluxes through consumption and excretion of compounds (Moslemi et al., 2012).

"Biological invasions are a growing problem worldwide" (van Wilgen et al., 2014). Researchers generally agree that invasive plant alien species represent "a major threat to biodiversity." Exotic plant species have the capacity to contribute to the extinction of local plant species. The present meta-analysis confirmed that invasive plant species contribute to the decline of native plant species and the local extinction of particular plant species (Gaertner et al., 2009).

Invasive plant species are common in South Africa. In addition to being fire prone, such species represent a "major threat" to biodiversity and water sources in South Africa (Gambrog et al., 2012).

Hundreds of the thousands of plant species that were

introduced to South Africa in the past 360 years have become invasive species. Invasive species can pose significant threat to the ecosystems of South Africa. They can increase erosion rates, reduce water supply, aggravate wildfires, contribute to the deterioration of rangelands, reduce agricultural productivity, pose threats to the health of human beings and livestock and impact biodiversity in a number of other ways (van Wilgen et al., 2014).

The ability of exotic plant species to invade native plant species is highly variable. Variability has been primarily attributed to the biotic resistance of plant communities. Whether natural or human-driven, wildfires affect the biodiversity of Mediterranean forests in the world. Alien species invasion is facilitated by the increase of fire intensity and severity. The recovery of native plant species following wildfires prevents the persistence of invasive plant species on burned areas (Pino et al., 2013).

Less than 10 countries worldwide have devised a screening process for newly emerging invasive plant species. It appears that "the majority of countries are stuck in the 'damage control' phase in which established weeds are being continuously contained or maintained, and very few have had the opportunity to try to eradicate newly arrived ('emerging') alien species before they have had the chance to become well established" (Lalla, 2014).

26.5.9 Mutualism

The interaction between ants and plants is a representation of the biology of mutualisms. While ants provide a protection to plants against herbivore threats, plants provide food and shelter for ants (Campbell et al., 2013). The soil interacts with plant species. Changes in plant composition contribute to alterations in soil components and vice versa (Ehrenfeld et al., 2005).

26.5.10 Protected (Conservation) Areas

Environmental protection and conservation legislation can result in "bureaucracy that has little practical meaning." "It is not uncommon to enact legislation or other rule-making that not only does not help the area or species of concern but may in fact impede or hinder resolution or recognition of a perceived problem. Such regulation may restrict scientific research while at the same time provide a false sense of achievement" (Shinn, 2004).

Worldwide protected areas have been growing at an accelerated rate in the past 25 years. Concurrently, however, "the mission of protected areas has expanded from biodiversity conservation to improving human welfare. The result is a shift in favor of protected areas allowing local resource use." At present, over 100,000 regions have been designated as protected areas, encompassing a total of 17.1 million km^2 or 11.5% of the global land surface. 25 years ago, protected areas were to the most part "the domain" of forestry officials, ecologists and occasionally, of terrestrial-use planners. As of today, protected areas are "included in the international arena as part of the Millennium Developmental Goals." The main objective of the UN Millennium Goal is to "eradicate extreme poverty and end hunger". Protected areas are currently "expected to directly contribute to national development and poverty reduction." Poverty alleviation initiatives promoted by agencies "devoted to this task" in the past half century have "uneven record" (Naughton-Treves et al., 2005).

"Protected areas are under threat in many parts of the world" (Ritchie et al., 2013). One third of the global terrestrial surface has been converted to agricultural or urban areas while an additional one third of land surface is projected to be transformed in the next 100 years (Naughton-Treves et al., 2005). Human population size is growing at an ecologically unsustainable pace. The human population accounted for 7 billion people between October 2011 and March 2012 and is expected to reach 7.5 to 8 billion by 2025, 11 billion by 2025 and 16 billion by 2100 (Webb, 2013). Population growth, civil conflict, increased consumption of resources, climate change, a doubling in food demand in the next 50 years as well as expansion of development activities are expected to intensify the "pressure on remaining natural habitats and the biodiversity within them." The majority of protected areas in the world "are open to some form of human use." From a total of 98,400 terrestrial protected areas in the world, 8,800 (8.9%) are listed under strict protection I and II categories of the International Union for Conservation of Nature (IUCN).

Terrestrial surfaces managed by indigenous people worldwide have often retained high levels of biodiversity. Energy and mineral exploration and large-scale development activities (i.e., infrastructure) are the major threats to protected and indigenous areas (Naughton-Treves et al., 2005).

26.5.10.1 Australia

The national reserve system in Australia is not ideal, yet it does protect some of the country's "unique species and ecosystems." Although Australia's terrestrial reserve system occupies 13.4% of Australia's landmass, "disproportionately small percentage of productive landscapes" is protected. Similarly, marine reserve areas are located in zones with the least opportunities for commercial fishing and fossil-fuel development. The integrity of Australia's terrestrial and marine reserve system is being compromised in spite of biodiversity losses.

There is a concern about policy, management and legislative changes being undertaken by three of six Australian governments "for exploitative uses of national parks." Recently proposed or enacted legislature will allow for "an increase in exploitative uses of reserves – including industrial logging, grazing by domestic livestock, mining, commercial development, and recreational hunting and fishing – all of which are detrimental to nature conservation." "Hard-won" laws limiting the clearing of vegetation on private properties are being "relaxed" by the governments of Queensland and Victoria which will inevitably contribute to additional loss of biodiversity. Despite the fact that 85% of high-conservation value vegetation is located on private land, the New South Wales government is deliberating regarding "relaxing anticlearing laws" on private land. "In Western Australia there have been large excisions of existing conservation land for mining." The Leadbeater's possum is listed as endangered by the International Union for Conservation of Nature. The "Victorian government is knowingly condoning activities that will reduce the viability of this species." Increased access to fishing and hunting in national reserves is also being considered by Australian governments. "Economic rationals are being used to justify the dismantling of park protections...The repair bill for these effects will dwarf any short-term economic benefits to extractive industries, and some changes might be irreversible."

"The fact that state governments are retreating from the previously accepted principal purpose of reserves-to conserve biodiversity-suggests a shortsighted decline in political and societal concern for nature conservation" (Ritchie et al, 2013).

26.5.10.2 Ecuador

The Manager of Machalilla National Park in Ecuador recently made a public promise that the 39,000 hectares protected area "would serve as a maquina de dinero (money machine) for the surrounding province" despite an Ecuadorian botanist statement that "the park represented the last hope for sustaining endemic species found in coastal dry forests" (Naughton-Treves et al, 2005).

26.5.10.3 Quebec

In 2012, Quebec Premier Jean Charest presented the commitment of his government to dedicate half of northern Quebec to biodiversity conservation. The biodiversity plan is part of Plan Nord which focuses on energy and mining development. It allocated 20% of the land as conservation area while prohibiting industry development in 30% of the area. Plan Nord is considered to be "one of the largest land-conservation initiatives in human history." Plan Nord "encompasses" approximately 1.2 million km² of land which is equivalent to 72% of Quebec's area, three times as large as the state of California in the U.S. The Plan Nord contains one of the world's largest freshwater reserves and 200,000 km² of commercial forests. Hundreds of millions of birds, herds and fish reside within the region. The area is further abundant in metals (i.e., gold; cobalt; nickel; iron ore; zinc; platinum) and is characterized by its wind-power potential and hydroelectricity derived from the region's rivers. The public and environmental groups, however, question the exclusion of industry from 30% of the area and fear that only regions that are not profitable for industries will remain as conservation areas by 2035. There is also a concern that mineral exploration and forest logging will be categorized as non-industrial initiatives. A number of governmental scientists further contend that "it is not a good strategy to focus conservation on industry exclusion" while it would be a "waste of management resources" to assign a "special status" to areas that have no "developmental potential" and are therefore "naturally protected." Currently, 120,000 people, including 33,000 aboriginal populations reside within the Plan Nord region (Berteaux, 2013).

26.6 Public

There is an increasing awareness that people "are at the heart of conservation." "People want to have a voice in decisions that affect them." Engaging people in decision-making processes ensures the establishment of effective democracy (Edwards & Gibeau, 2013).

"Scholars have recognized the importance of citizen involvement in a democratic government for thousands of years.". The First Amendment of the U.S. Constitution affords the right of U.S. citizens to petition the government. Citizen participation in governments has been "crucial" in establishing, designing, and enforcing environmental laws. Strategic lawsuit against public participation (SLAPP) is a form of retaliation against citizen participation in governments that "interfere with a party's exercise of its right to petition the government, typically by draining the party's time and resources." While a number of states have implemented anti-SLAPP protection measures, such protections are not available on the level of the federal government. "There is currently no effective defense against a SLAPP" (O'Neill, 2011).

***

"Each of us can contribute to a healthier home for all of us on planet Earth – just by making a better choice" (Kreisberg, 2007).

Discussion

Within the context of reviewed research, "Planet De-Construction" allows us to draw a couple of conclusions relative to Sosé's theory.

First, Ecuador and the Australian government comrpomising protected areas (i.e., problem) in the name of profit and EPA's "Take Permits" are yet another clear indication that financial gain is a top priority for governments.

Similarly to all thus far discussed governmental agencies (i.e., U.S. Marine Mammal Protection Act; CAA), EPA is acting upon the exact opposite principles to the ones it publicly stands for. Its very purpose is to protect species as the name of the agency implies, yet instead, it threatens and endangers species' continual existence (i.e., deceit and problem) in the name of profit. The "Take Permits" in addition to the failure of recovery plans and ambiguity of EPA's terminology and regulations ensure the continual flow of funds to the agency through created problems and donor funds. Recovered species automatically translate into cessation of funding. There would be no need for funding recovery plans if species were healthy and abundant. Only endangered species generate profit for the EPA. The larger the problems, the greater the profit.

Second, "Planet De-Construction" provides yet another clear example of how problems are created for purposes of profit. Protected areas are presently being utilized as resource-exploitation sites to alleviate poverty in the world. Yet, as was made apparent throughout "Sovereign Terra," poverty is intentionally created (i.e., problem) by governments and industries, leading to the acquisition of governmental and industrial profit and power.

References

Akkermans, T., Thiery, W., Van Lipzig, N. P. M. (2014). The regional climate impact of a realistic future deforestation scenario in the Congo Basin. *Journal of Climate, 27*, 2714-2734.

Bader, M. K.-F., Leuzinger, S., Kell, S. G., Siegwolf, R. T. W., Hagedorn, F., Schleppi, P. & Korner, C. (2013). Central European hardwood trees in a high-CO2 future: Synthesis of an 8-year forest canopy CO2 enrichment project. *Journal of Ecology, 101*, 1509-1519.

Bagley, J. E., Desai, A. R., Harding, K. J., Snyder, P. K. & Foley, J. A. (2014). Drought and deforestation: Has land cover change influenced recent precipitation extremes in the Amazon? *Journal of Climate, 27,* 345-361.

Bennett, J. R. & Arcese, P. (2013). Human influence and classical biogeographic predictors of rare species occurrence. *Conservation Biology, 27*(2), 417-421.

Berteaux, D. (2013). Quebec's large-scale Plan Nord. *Conservation Biology, 27*(2), 242-247.

Bitariho, R. & McNeilage, A. (2007). Population structure of montane bamboo and causes of its decline in Echuya Central Forest Reserve, South West Uganda. *African Journal of Ecology, 46,* 325-332.

Bourgeon, G., Nair, K. M., Ramesh, B. R. & Seen, D. L. (2012). Consequences of underestimating ancient deforestation in South India for global assessments of climatic change. *Current Science, 102*(12), 1699-1703.

Bradshaw, C. J. A., Sodhi, N. S., Pen, K. S.-H. & Brook, B. W. (2007). Global evidence that deforestation amplifies flood risk and severity in the developing world. *Global Change Biology, 13,* 2379-2395.

Carroll, C., Vucetich, J. A., Nelson, M. P., Rohlf, D. J. & Phillips, M. K. (2010). Geography and recovery under the U.S. Endangered Species Act. *Conservation Biology, 24*(2), 395-403.

Chen, Y., Randerson, J. T., van der Werf, G., Morton, D. C., Mu, M. & Kasibhatla, P.S. (2010). Nitrogen deposition in tropical forests from savanna and deforestation fires. *Global Change Biology, 16,* 2024-2038.

Cheng, J. (2009). Court ordered listings: *Center for Biological Diversity v. Fish and Game Commission. Ecological Law Quarterly, 36,* 569-574.

Collen, B., Purvis, A. & Mace, G. M. (2010). When is a species really extinct? Testing extinction inference from a sighting record to inform conservation assessment. *Diversity and Distributions, 16,* 755-764.

Conceicao, P. & Mendoza, R. U. (2009). Anatomy of the global food crisis. *Third World Quarterly, 30,* 1159-1182.

Crafford, J. G. & Hassan, R. M. (2014). Relationships between ecological infrastructure and the economy: The case of a fishery. *South African Journal of Science*, *110*(7/8), 43-50.

Craft, C. B. (2012). Tidal freshwater forest accretion does not keep pace with sea level rise. *Global Change Biology*, *18*, 3615-3623.

Crous, C. J. & Roets, F. (2014). Realising the value of continuous monitoring programmes for biodiversity conservation. *South African Journal of Science*, *110*(11/12), 7-11.

Davin, E. L. & de Noblet-Ducoudre, N. (2010). Climatic impact of global-scale deforestation: Radiative versus nonradiative processes. *Journal of Climate*, *23*, 97-112.

Doak, D. F., Bakker, V. J. & Vickers, W. (2013). Using population viability criteria to assess strategies to minimize disease threats for an endangered carnivore. *Conservation Biology*, *27*(2), 303-314.

Dunn, R. R., Harris, N. C., Colwell, R. K., Koh, L. P. & Sodh, N. S. (2009). The sixth mass coextinction: Are most endangered species parasites and mutualists? *Proceedings of the Royal Society B*, *276*, 3037-3045.

Edwards, F. N. & Gibeau, M. L. (2013). Engaging people in meaningful problem solving. *Conservation Biology*, *27*(2), 239-241.

Ehrenfeld, J. G., Ravit, B. & Elgersma, K. (2005). Feedback in the plant-soil system. *Annual Review of Environment & Resources*, *30*, 75-115.

Ethier, D. M., Lafleche, A., Swanson, B. J. Nocera, J. J. & Kyle, C. J. (2012). Population subdivision and peripheral isolation in American badgers (Taxidea taxus) and implications for conservation planning in Canada. *Canadian Journal of Zoology*, *90*, 630-639.

Faith, J. T. & Thompson, J. C. (2013). Fossil evidence for seasonal calving. *Journal of Biogeography*, *40*, 2108-2118.

Feeley, K. J., Malh, Y., Zelazowki, P. & Silman, M. R. (2012). The relative importance of deforestation, precipitation change, and temperature sensitivity in determining the future distributions and diversity of Amazonian plant species. *Global Change Biology, 18*, 2636- 2647.

Fennesy, J., Bok, F., Tutchings, A., Brenneman, R. & Janke, A. (2013) Mitochondrial DNA analyses show that Zambia's South Luangwa Valley giraffe (Giraffa Camelopardalis thornicrofti) are genetically isolated. *African Journal of Ecology, 51*, 635-640.

Finseth, R. M. & Conrad, J. M. (2014). Cost-effective recovery of an endangered species: The red-cockaded woodpecker. *Land Economics, 90*(4), 649-667.

Fisaha, G., Hundera, K. & Dalle, G. (2013). Woody plants' diversity, structural analysis and regeneration status of Wof Washa natural forest, North-east Ethiopia. *African Journal of Ecology, 51*, 599-608.

Folke, C., Hahn, T., Olsson, P. & Norberg, J. (2005). Adaptive governance of social-ecological systems. *Annual Review of Environment & Resources, 30*, 441-473.

Gambrog, C., Millar, K., Shortall, O. & Sandoe, P. (2012). Bioenergy and land use: Framing the ethical debate. *Journal of Agricultural & Environmental Ethics, 25*, 909-925.

Gaertner, M., Breeyen, A. D., Hui, C. & Richardson, D. M. (2009). Impacts of alien plant invasions on species richness in Mediterranean-type ecosystems: A meta-analysis. *Progress in Physical Geography, 33*(3), 319-338.

Gersen, S. (2009). Who can enforce the Endangered Species Act's command for federal agencies to carry out conservation programs? *Ecology Law Quarterly, 36*, 407-438.

Gregory, R., Arvai, J. & Gerber, L. R. (2013). Structuring decisions for managing threatened and endangered species in a changing climate. *Conservation Biology, 27*(6), 1212-1221.

Hahn, M. B., Gangnon, R. E., Barcellos, C., Asner, G. P. & Patz, J.A. (2014). Influence of deforestation, logging, and fire on malaria in the Brazilian Amazon. *PLoS ONE, 9*(1), 1-8.

Hammerly, S. C., Morrow, M. & On, J. J. (2013). A comparison of pedigree-and DNA-based measures for identifying inbreeding depression in the critically endangered Attwater's Prairie-chicken. *Molecular Ecology, 22*, 5313-5328.

Hanspach, J., Fischer, J., Ikin, K., Stott, J. & Law, B. S. (2012). Using trait-based filtering as a predictive framework for conservation: A case study of bats on farms in southeastern Australia. *Journal of Applied Ecology, 49*, 842-850.

Hartmann, H., Ziegler, W., Kolle, O. & Trumbore, S. (2013). Thirst beats hunger – declining hydration during drought prevents carbon starvation in Norway spruce saplings. *New Phytologist, 200*, 340-349.

Horrocks, M., Salter, J., Braggins, J. Nichol, S., Moorhouse, R. & Elliotte, G. (2008). Plant microfossil analysis of coprolites of the critically endangered kakapo (Strigops habroptilus) parrot from New Zealand. *Review of Palaeobotany and Palynology, 149*, 229 -245.

Hughes, T. P., Huang, H. & Young, M. A. L. (2012). The wicked problem of China's disappearing coral reefs. *Conservation Biology, 27*(2), 261-269.

Imura, H., Yedla, S., Shirakawa, H. & Memon, M. A. (2005). Urban environmental issues and trends in Asia-an overview. *International Review for Environmental Strategies, 5*(2), 357-382.

Johansson, O., Oalmqvist, K. & Olofsson, J. (2012). Nitrogen deposition drives lichen community changes through differential species responses. *Global Change Biology, 18*, 2626-2635.

Johnson, S. L., Kuske, C. R., Carney, T. D., Housman, D. C., Gallegos-Graves, La V. & Belnap, J. (2012). Increased temperature and altered summer precipitation have differential effects on biological soil crusts in a dryland ecosystem. *Global Change Biology, 18*, 2583-2593.

Juslen, A., Hyvarinen, E. & Virtanen, L. K. (2013). Application of the red-list index at a national level for multiple species groups. *Conservation Biology, 27*(2), 398-406.

Kartzinel, T. R., Trapnell, D. W. & Shefferson, R. P. (2013). Critical importance of large native trees for conservation of a rare Neotropical epiphyte. *Journal of Ecology, 101*, 1429-1438.

Kirch, P. V. (2005). Archeology and global change: The Holocene record. *Annual Review of Environment & Resources, 30*, 409-440.

Klopper, R. R., Gautier, L., Smith, G. F., Spichiger, R. & Chatelain, C. (2006). Inventory of the African flora: A world first for the forgotten continent. *South African Journal of Science, 102*, 185-186.

Kreisberg, J. (2007). Greener pharmacy: Proper medicine disposal protects the environment. *Integrative Medicine, 6*(4), 50-52.

Krishna, P. H., Saranya, K. R. L., Reddy, C. S., Jha, C. S. & Dadhwal, V. K. (2014). Assessment and monitoring of deforestation from 1930 to 2011 in Andhra Pradesh, India using remote sensing and collateral data. *Current Science, 107*(5), 867-875.

Krishnadas, M., Nair, T. & Karnad, D. (2013). Equality in conservation: Comment on Bawa et al. 2011. *Conservation Biology, 27*(2), 422-424.

Laikre, L., Jansson, M., Allendore, F. W., Jakobsson, S. & Ryman, N. (2012). Hunting effects on favourable conservation status of highly inbred Swedish wolves. *Conservation Biology, 27*(2), 248-253.

Lalla, R. (2014). South Africa – A global player in the battle against alien plant invasions. *South African Journal of Science, 110*(7/8), 4-6.

Leidner, A. K. & Neel, M. C. (2011). Taxonomic and geographic patterns of decline for threatened and endangered species in the United States. *Conservation Biology, 25*(4), 716-725.

Lewis, M. B. & Schupp, E. W. (2014). Reproductive ecology of the endangered Utah endemic Hesperidanthus suffrutescens with implications for conservation. *American Midland Naturalist, 172*, 236-251.

Loehle, C. & Eschenbach, W. (2012). Historical bird and terrestrial mammal extinction rates and causes. *Diversity and Distributions, 18*, 84-91.

Lohmann, D., Tietjen, B., Blaum, N., Joubert, D. F. & Jeltsch, F. (2012). Shifting thresholds and changing degradation patterns: Climate change effects on the simulated long-term response of a semi-arid savanna to grazing. *Journal of Applied Ecology, 49*, 814-823.

Marquard-Peterson, U. (2012). Decline and extermination of an Arctic wold population in East Greenland, 1899 – 1939. *Arctic, 65*(2), 155-166.

Mauderly, J. L. & Chow, J. C. (2008). Health Effects of Organic Aerosols. *Inhalation Toxicology, 20*, 257-288.

McClure, M. M., Alexander, M., Borggaard, D., Boughton, D., Crozier, L., Griffis, R., Jorgensen, J. C., Lindley, S. T., Nye, J., Rowland, M. J., Seney, E. E., Snover, A., Toole, C. & McWethy, D., Whitlock, C., Wilmshurst, J. M., McGlone, M. S. & Li, X. (2009). Rapid deforestation of South Island, New Zealand, by early Polynesian fires. *The Holocene, 19*(6), 883-897.

Medvigy, D., Walko, R. L., Otte, M. J. & Avissar, R. A. (2013). Simulated changes in Northwest U.S. climate in response to Amazon deforestation. *Journal of Climate, 26*, 9115-9136.

Memon, M. A., Pearson, C. & Imura, H. (2005). Inter-city environmental cooperation: The case of the Kitakyushu initiative for a clean environment. *International Review for Environmental Strategies, 5*(2), 531-540.

Moslemi, J. M., Snider, S. B., MacNeill, K., Gilliam, J. F. & Flecker, A. S. (2012). Impacts of an invasive snail (*Tarebia granifera*) on nutrient cycling in tropical streams: The role of riparian deforestation in Trinidad, West Indies. *PLoS ONE*, *7*(6), 1-9.

Nagamitsu, T., Kikuchi, S., Hotta, M., Kenta, T. & Hiura, T. (2014). Effects of population size, forest fragmentation, and urbanization on seed production and gene flow in an endangered maple (*Acer miyabei*). *American Midland Naturalist*, *172*, 303-316.

Naughton-Treves, L., Holland, M. B. & Brandon, K. (2005). The role of protected areas in conserving biodiversity and sustaining local livelihoods. *Annual Review of Environment & Resources*, *30*, 219-252.

Nepstad, D., Schwartzman, S., Bamberger, B., Santilli, M., Ray, D., Schlesinger, P., Lefebvre, P., Alencar, A., Prinz, E., Fiske, G. & Rolla, A. (2006). Inhibition of Amazon deforestation and fire by parks and indigenous lands. *Conservation Biology*, *20*(1), 65-73.

O'Keeffe, J. (2009). Sustaining river ecosystems: Balancing use and protection. *Progress in Physical Geography*, *33*(3), 339-357.

O'Neill, J. J. (2011). The Citizen Participation Act of 2009: Federal legislation as an effective defense against SLAPPs. *Environmental Affairs*, *38*, 477-507.

Ouvrier, S. J., Holland, M., Stroever, J., Barbraud, C., Weimerskirch, H., Serreze, M. & Caswell, H. (2012). Effects of climate change on an emperor penguin population: Analysis of coupled demographic and climate models. *Global Change Biology, 18*, 2756-2770.

Packman, C. E., Gray, T. N. E., Collar, N. J., Evans, T. D., Van Zalinge, R. N., Virak, S., Lovett, A. A. & Dolman, P. M. (2013). Rapid loss of Cambodia's grasslands. *Conservation Biology, 27*(2), 245-247.

Pagano, A. M., Durner, G. M., Amstrup, S. C., Simac, K. S. & York, G. S. (2012). Long-distance swimming by polar bears (Ursus maritimus) of the southern Beaufort Sea during years of extensive open water. *Candian Journal of Zoology, 90,* 663-676.

Paeth, H., Born, K., Girmes, R., Podzun, R. & Jacob, D. (2009). Regional climate change in tropical and Northern Africa due to greenhouse forcing and land use changes. *Journal of Climate, 22,* 114-132.

Payne-Sturges, D. & Kemp, D. (2008). Ten years of addressing children's health through regulatory policy at the U.S. Environmental Protection Agency. *Environmental Health Perspectives, 116*(12), 1720-1724.

Pino, J., Arnan, X., Rodrigo, A. & Retana, J. (2013). Post-fire invasion and subsequent extinction of *Conyza* spp. in Mediterranean forests is mostly explained by local factors. *European Weed Research Society, 53,* 470-478.

Povilitis, A. & Suckling, K. (2010). Addressing climate change threats to endangered species in U.S. recovery plans. *Conservation Biology, 24*(2), 372-376.

Regan, T. J., Taylor, B. L., Thompson, G. G., Cochrane, J. F., Ralls, K., Runge, M. C. & Merrick, R. (2013). Testing decision rules for categorizing species' extinction risk to help develop quantitative listing criteria for the U.S. Endangered Species Act. *Conservation Biology, 27*(4), 821-831.

Ribon, R., Simon, J. E., De Mattos, G. T. (2003). Bird extinctions in Atlantic forest fragments of the Vicosa Region, Southeastern Brazil. *Conservation Biology, 17*(6), 1827-1839.

Ritchie, E. G., Bradshaw, C. J. A., Dickman, C. R., Hobss, R., Johnson, C. N., Johnson, E. L., Laurance, W. F., Lindenmayer, D., McCarthy, M. A., Nimmo, D. G., Possingham, H. H., Pressey, R. L., Watson, D. M. & Woinarski, J. (2013). Continental-scale governance and the hastening of loss of Australia's biodiversity. *Conservation Biology, 27*(6), 1133- 1135.

Rius, D., Vanniere, B. & Galop, D. (2009). Fire frequency and landscape management in the northwestern Pyrenean piedmont, France, since the early Neolithic (8000 cal. BP). *The Holocene, 19*(6), 847-859.

Robertson, T. (2012). Total war and the total environment: Fairfield Osborn, William Vogt, and the birth of global ecology. *Environmental History, 17,* 336-364.

Sajeva, M., Augugliaro, C., Smith, M. J. & Oddo, E. (2013). Regulating internet trade in CITES species. *Conservation Biology, 27*(2), 429-430.

Schosler, H., de Boer, J. & Boersema, J. J. (2013). The organic food philosophy: A qualitative exploration of the practices, values, and beliefs of Dutch organic consumers within a cultural-historical frame. *Journal of Agricultural & Environmental Ethics, 26,* 439-460.

Shinn, E. A. (2004). The mixed value of environmental regulations: Do acroporid corals deserve endangered species status? *Marine Pollution Bulletin, 49,* 531-533.

Sollomann, R., Gardner, B., Chandler, R. B., Shindle, D. B., Onorato, D. P., Royle, J. A. & O'Connell, A. F. (2013). Using multiple data sources provides density estimates for endangered Florida panther. *Journal of Applied Ecology, 50,* 961-968.

Staub, C. G., Binford, M. W. & Stevens, F. R. (2013). Elephant herbivory in Majete Wildlife Reserve, Malawi. *African Journal of Ecology, 51,* 536-543.

Stirling, I. & Derocher, A. E. (2012). Effects of climate warming on polar bears: A review of the evidence. *Global Change Biology, 18,* 2694-2706.

van Wilgen, B. W., Davies, S. J. & Richardson, D. M. (2014). Invasion science for society: A decade of contributions from the Centre for Invasion Biology. *South African Journal of Science, 110*(7/8), 8-19.

Vukovich, M. & Ritchison, G. (2008). Foraging behavior of short-eared owls and Northern harriers on a reclaimed surface mine in Kentucky. *Southern Naturalist, 7*(1), 1-10.

Wang, Y. & Zhou, J. (2013). Endocrine disrupting chemicals in aquatic environments: A potential reason for organism extinction? *Aquatic Ecosystem Health & Management*, *16*(1), 88-93.

Webb, E. C. (2013). The ethics of meat production and quality - a South African perspective. *South African Journal of Animal Science*, *43*(5), S2-S10.

Welch, J. R., Brondizio, E. S., Hetrick, S. S. & Coimbra Jr, C. E. A. (2013). Indigenous burning as conservation practice: Neotropical savanna recovery amid agribusiness deforestation in Central Brazil. *PLoS ONE*, *8*(12), 1-10.

Wetzel, F., Beissmann, H., Penn, D. J. & Jetz, W. (2013). Vulnerability of terrestrial island vertebrates to projected sea-level rise. *Global Change Biology*, *19*, 2058-2070.

Wittmer, H. U., Serruoya, R., Elbroch, L. M. & Marshall, A. J. (2012). Conservation strategies for species affected by apparent competition. *Conservation Biology*, *27*(2), 254-260.

Yonekura, Y., Ohta, S., Kiyono, Y., Aksa, D., Morisada, K., Tanaka, N. & Tayasu, I. (2012). Dynamics of soil carbon following destruction of tropical rainforest and the subsequent establishment of Imperata grassland in Indonesian Borneo using stable carbon isotopes. *Global Change Biology*, *18*, 2606-2616.

Yuknis, E. M. (2011). Would a "God Squad" exemption under the Endangered Species Act solve the California water crisis? *Boston College Environmental Affairs Law Review*, *38*, 567-596.

Zeigler, S. L., Che-Castaldo, J. P. & Neel, M. C. (2013). Actual and potential use of population viability analyses in recovery of plant species listed under the U.S. Endangered Species Act. *Conservation Biology*, *27*(6), 1265-1278.

CONCLUSION

"Sovereign Terra" proposes a theory devised by Sosé Gjelaj which states that problems are intentionally created by industries and governments for purposes of generating financial gain and power. Governments and industries are one and the same entity (i.e., oil industry, etc.) with the same purpose (i.e., financial gain and power), disguised under two different names and functions. The greater the problems, the larger the profit and acquired power (i.e., food aid sector, etc.).

A number of examples were elaborated upon in support of Sosé's theory (i.e., World War II giving rise to GM agriculture which in return increases profit and power for GM companies; industrial and chemical substances creating disease from which pharmaceutical companies and health care sectors profit; no access to food birthing profit and power to food aid organizations, etc.).

To acquire profit and power, truth is manipulated and hidden from humanity (i.e., secrecy on the levels of GMOs, nano technology, food aid organizations, packaged salad greens, etc.) while legal doors (i.e., GMO, nano technologies substantial equivalence biosafety standards; EPA's "Take Permit" and ambiguous terminology; no geoengineering regulations, etc.) are opened to allow for profit and power to be generated. The deceit is exemplified by governments creating the illusion that problems are attended to by governmental regulations, organizations (i.e., Environmental Protection Agency; U.S. Marine Mammal Protection Act; Clean Air Act; Clean Water Act; Food Aid; Judicial System, etc.) and actions (i.e., biofuels to replace fossil fuel; electrical cars to substitute traditional light vehicles, etc.) when in fact problems are perpetuated by such governmental entities and endeavors. The same deceiving, problem-profit strategies became apparent across all sectors (i.e., agricultural; air;

415

water).

The largest deliberately orchestrated problems appear to be economic inequality, human and environmental destruction. All of them lead to the concentration of wealth and power in the hands of the few (i.e., GM companies). The established problem-profit-power trajectory leads to the conclusion that the end goal of governments and industries is acquiring absolute ownership of Earth's resources, population control (i.e., reproductive system impairments) and absolute power over humanity. The "Corporal Violations" example related to the expensive and effective water treatment systems available in certain parts of the world provides an indication that the few who were gaining momentum in absolute ownership of Earth had the means to survive the environmental destruction inflicted upon the rest of human kind.

Sosé's theory ends by stating that the awareness of the truth is sufficient for humanity to find real solutions to root problems, environment and human-wise. Examples providing support for such aspect of Sosé's theory are apparent (i.e., the public mistrusting GMO companies due to the awareness of GMOs detrimental human and environmental impact; countries refusing GM crops as food donations, etc.).

Having had acquired vital knowledge and understanding from "Sovereign Terra" in regards to the true colors of reality, what actions are we, individually and as a collective consciousness, going to partake upon? The number of examples on interconnectedness (i.e., mass extinction) suggest that all humanity is inevitably going to be severely affected by governmental and industrial agendas. Earth and humanity are in a state of emergency, on the verge of a global collapse, on the frontier of extinction. Are we going to wait for time to engulf us into its innate limitation or are we going to fast-forward it to the moment of realization of the inner power residing within us and take action to save ourselves and our planetary home? Do we dare to live, thrive, freely

416

express who we truly are, heal the environment and expand as a collective consciousness? Are we open to the possibility of being abundant, healthy, happy and free, embracing the energy of love within our hearts and projecting it outward? Are we ready to create the reality that we wish to experience?

We are!

ABOUT THE AUTHORS

Sosé Gjelaj was born in Montenegro. She relocated to the United States with her family in 1969 and has lived in America since then. Sosé is an author, a philanthropist, a published poet and a pronoun artist who had spent a lifetime studying Eastern and Western philosophy. She has academic background in Arts and is the owner of "Sosé Art Gallery" located in Bennington, VT. Sosé is the founder of the "Source of Visibility," a not-for-profit humanitarian and environmental organization.

Elitsa Teneva was born in Bulgaria. She moved to the United States when she was 18 to pursue an academic career in the fields of psychology and school psychology. She graduated with a B.S. in Psychology and a minor in Child Development from Southern Vermont College in Bennington, VT, and M.Ed. with a concentration in School Psychology from the University of Massachusetts Amherst, Amherst, MA. Elitsa is a member on the Board of Directors of the "Source of Visibility".

OTHER BOOKS BY THE AUTHORS

NOUN=VERB by Sosé Gjelaj and Elitsa Teneva

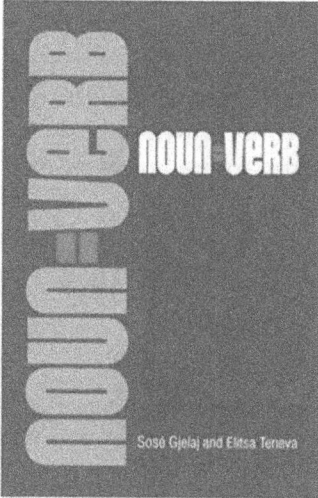

Based on over 250 research articles, Noun=Verb explores the fundamental reasons for the staggering rates of psychological disorders, violence, incarceration, failure to graduate, unemployment, suicide and exploitation among humanity. Findings demonstrate that maladjustment and trauma are not accidental. Rather, such events are deliberately orchestrated by various governmental sectors (i.e., educational system; public and mental health; judicial system) relying on an array of harmful methods (i.e., fear appeals; punishment). The goal is the acquisition of absolute corporate profit at the expense of life. Solving the injustice would require awareness and the courage to cultivate and implement a value system founded on spirituality (i.e., focus on creativity; love; autonomy; reaching one's maximum potential; positive emotions) and not on self-serving corporate interest.

SYNCHRONIZATION OF DIMENSIONS by Sosé Gjelaj and Elitsa Teneva

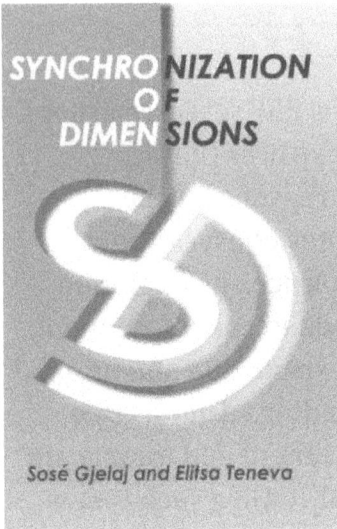

Synchronization of Dimensions takes us on a journey of the deepest questions and answers facing mankind today. Topics such as human and universally created laws, sexual energy and the difference between darkness and light are a few of the topics explored. Delving into the magic of wisdom, the reader is left with a clear understanding of his origins, spiritual makeup and genetic future.

THE 13TH SECRET CODE by Sosé Gjelaj

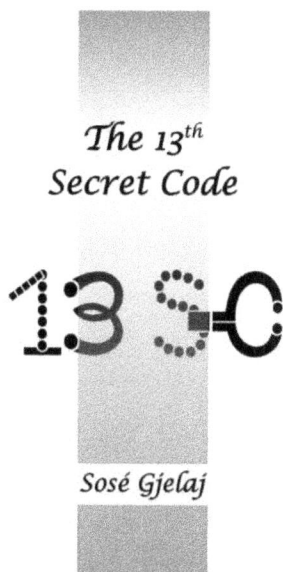

The 13th
Secret Code

Sosé Gjelaj

The 13th Secret Code is a compilation of selected poems written by Sosé Gjelaj in the past almost five decades. Sosé's wisdom and creativity flow through the pages as the reader delves into the ocean of source energy and the experiences of our eternal soul. Love, the evolution of human consciousness and the unknown are some of the topics that Sosé explores through her artistic pen. The reader is gifted with a masterpiece of mystery and magic.